Economic Development : Agricultural and Industrial Growth

Economic Development : Agricultural and Industrial Growth

D.R. Joshi
Sandeep Bajpai

CYBER TECH PUBLICATIONS
4264/3, Ansari Road, Daryaganj, New Delhi-110002 (India)
Ph.: 011-23244078, 011-43559448 Fax: 011-23280028
E-Mail: cyberpublicationsdelhi@yahoo.com
Website: www.cybertechpublications.com

ECONOMIC DEVELOPMENT : AGRICULTURAL AND INDUSTRIAL GROWTH

D.R. Joshi
Sandeep Bajpai

First Edition 2010

Published by Asian Offset Printers
G.S. Rawat for **Cyber Tech Publications**
4264/3, Ansari Road, Daryaganj, New Delhi-110002 (India)
Ph.: 011-23244078, 011-43559448 Fax: 011-23280028
E-Mail: cyberpublicationsdelhi@yahoo.com
Website: www.cybertechpublications.com

Preface

While analysing the mode of production in Indian agriculture it is necessary to do so based on the laws of capitalist production as discovered by Marx and further developed by Lenin. These scientific laws of capitalism and imperialism are not time bound. They neither get out-dated nor change from country to country. The conditions may change in time and space but the basic laws of capitalism and imperialism remain unchanged. To understand the mode of production in India it is no use negating these laws, as the CRC does, but utilise them to understand the mode of produc-tion in India.

Regarding the question of capitalism in agriculture an important principle outlined by Lenin was that, "it is the development of capitalism in the manufacturing industry that is the main force which gives rise to, and develops, capitalism in agriculture."

The growth of commodity production in agriculture can be assessed by the extent of utilisation of inputs and the growth of outputs produced for the market. The 'green revolution', which was introduced by the World Bank in Third World countries in the mid-1960s, was part of the imperialist policy to penetrate the coun-tryside for markets. This 'green revolution' has led to certain changes in agriculture which must be analysed. This section will be classified under the following sub heads : 1) 'Green Revolution', 2) Inputs, 3) Credit, 4) Output and Productivity Trends, and 5) Utilisation of Surplus

— *Author*

CONTENTS

1

INTRODUCTION

Agriculture has always been India's most important economic sector. Large area yield forecasting prior to harvest is of interest to government agencies, commodity firms and producers. Timely monitoring during the growing season provides opportunities to mitigate any detected challenges. The current methods used in India provide neither sufficient nor timely cost effective spatial information. The purpose of this project is to develop an Agricultural GeoCapacity Centre Network to overcome these problems and provide a sustainable, long term solution. This presentation describes the functionalities of AGCN, the requirements, the design and the implementation AGCN is a complex system where several independent modules are integrated and cooperate to reach the final objective. It is based on a multi-tier architecture model. These modules are implemented as one or several applications; for crop monitoring, yield forecasting, weather forecast, etc. The applications are built around services provided by the GIS core, like Image Processing, Map Generation, Web System, and Database Connection. The GIS core provides application programming interface to build custom applications, integrates external application and access to the database management system.

INTRODUCTION

Agriculture in India has a long history dating back to ten thousand years. Today, India ranks second worldwide in farm output. Agriculture and allied sectors like forestry, logging and fishing accounted for 16.6% of the GDP in 2007, employed 60%

of the total workforce and despite a steady decline of its share in the GDP, is still the largest economic sector and plays a significant role in the overall socio-economic development of India.

India is the largest producer in the world of milk, cashew nuts, coconuts, tea, ginger, turmeric and black pepper. It also has the world's largest cattle population (193 million). It is the second largest producer of wheat, rice, sugar, groundnut and inland fish. It is the third largest producer of tobacco. India accounts for 10% of the world fruit production with first rank in the production of banana and sapota.

India has many farm insurance companies that insure wheat, fruit, rice and rubber farmers in the event of natural disasters or catastrophic crop failure, under the supervision of the Ministry of Agriculture. One notable company that provides all of these insurance policies is agriculture insurance company of india and it alone insures almost 20 million farmers.

India's population is growing faster than its ability to produce rice and wheat. The most important structural reform for self-sufficiency is the ITC Limited plan to connect 20,000 villages to the Internet by 2013. This will provide farmers with up to date crop prices for the first time, which should minimise losses incurred from neighbouring producers selling early and in turn facilitate investment in rural areas.

Overview

India ranks second worldwide in farm output. Agriculture and allied sectors like forestry, logging and fishing accounted for 18% of the GDP in 2007, employed 60% of the total workforce and despite a steady decline of its share in the GDP, is still the largest economic sector and plays a significant role in the overall socio-economic development of India. Yields per unit area of all crops have grown since 1950, due to the special emphasis placed on agriculture in the five-year plans and steady improvements in irrigation, technology, application of modern agricultural practices and provision of agricultural credit and subsidies since Green revolution in India. However, international comparisons reveal that the average yield in India is generally 30% to 50% of the highest average yield in the world.

Indian Punjab is called the "Granary of India" or "India's bread-basket." It produces 14% of India's cotton, 20% of India's wheat, and 9% of India's rice.

Insurance

India has many farm insurance companies that insure wheat, fruit, rice and rubber farmers in the event of natural disasters or catastrophic crop failure, under the supervision of the Ministry of Agriculture. One notable company that provides all of these insurance policies is agriculture insurance company of india and it alone insures almost 20 million farmers.

Initiatives

The required level of investment for the development of marketing, storage and cold storage infrastructure is estimated to be huge. The government has implemented various schemes to raise investment in marketing infrastructure. Among these schemes are *Construction of Rural Go downs, Market Research and Information Network,* and *Development / Strengthening of Agricultural Marketing Infrastructure, Grading and Standardization.*

The Indian Agricultural Research Institute (IARI), established in 1905, was responsible for the research leading to the "Indian Green Revolution" of the 1970s. The Indian Council of Agricultural Research (ICAR) is the apex body in agriculture and related allied fields, including research and education. The Union Minister of Agriculture is the President of the ICAR.

The Indian Agricultural Statistics Research Institute develops new techniques for the design of agricultural experiments, analyses data in agriculture, and specializes in statistical techniques for animal and plant breeding. Prof. M.S. Swaminathan is known as "Father of the Green Revolution" and heads the MS Swaminathan Research Foundation. He is known for his advocacy of environmentally sustainable agriculture and sustainable food security.

Problems

Slow agricultural growth is a concern for policymakers as some two-thirds of India's people depend on rural employment for a living. Current agricultural practices are neither economically nor environmentally sustainable and India's yields for many agricultural commodities are low. Poorly maintained irrigation systems and almost universal lack of good extension services are among the factors responsible. Farmers' access to markets is hampered by poor roads, rudimentary market infrastructure, and excessive regulation.

—World Bank: "India Country Overview 2008"

The low productivity in India is a result of the following factors:

- According to World Bank's "India: Priorities for Agriculture and Rural Development", India's large agricultural subsidies are hampering productivity-enhancing investment. Overregulation of agriculture has increased costs, price risks and uncertainty. Government intervenes in labor, land, and credit markets. India has inadequate infrastructure and services. World Bank also says that the allocation of water is inefficient, unsustainable and inequitable. The irrigation infrastructure is deteriorating.
- Illiteracy, general socio-economic backwardness, slow progress in implementing land reforms and inadequate or inefficient finance and marketing services for farm produce.
- The average size of land holdings is very small (less than 20,000 m^2) and is subject to fragmentation, due to land ceiling acts and in some cases, family disputes. Such small holdings are often over-manned, resulting in disguised unemployment and low productivity of labour.
- Adoption of modern agricultural practices and use of technology is inadequate, hampered by ignorance of such practices, high costs and impracticality in the case of small land holdings.

- Irrigation facilities are inadequate, as revealed by the fact that only 52.6% of the land was irrigated in 2003–04, which result in farmers still being dependent on rainfall, specifically the Monsoon season. A good monsoon results in a robust growth for the economy as a whole, while a poor monsoon leads to a sluggish growth. Farm credit is regulated by NABARD, which is the statutory apex agent for rural development in the subcontinent.

History

Indian agriculture began by 9000 BCE as a result of early cultivation of plants, and domestication of crops and animals. Settled life soon followed with implements and techniques being developed for agriculture. Double monsoons led to two harvests being reaped in one year. Indian products soon reached the world via existing trading networks and foreign crops were introduced to India. Plants and animals—considered essential to their survival by the Indians—came to be worshiped and venerated.

The middle ages saw irrigation channels reach a new level of sophistication in India and Indian crops affecting the economies of other regions of the world under Islamic patronage. Land and water management systems were developed with an aim of providing uniform growth. Despite some stagnation during the later modern era the independent Republic of India was able to develop a comprehensive agricultural program.

AGRICULTURE

Agriculture refers to the production of food and goods through farming and forestry. Agriculture was the key development that led to the rise of civilization, with the husbandry of domesticated animals and plants (i.e. crops) creating food surpluses that enabled the development of more densely populated and stratified societies. The study of agriculture is known as agricultural science (the related practice of gardening is studied in horticulture).

Agriculture encompasses a wide variety of specialties. Cultivation of crops on arable land and the pastoral herding of livestock on rangeland remain at the foundation of agriculture. In the past century a distinction has been made between sustainable agriculture and intensive farming. Modern agronomy, plant breeding, pesticides and fertilizers, and technological improvements have sharply increased yields from cultivation. Selective breeding and modern practices in animal husbandry such as intensive pig farming (and similar practices applied to the chicken) have similarly increased the output of meat. The more exotic varieties of agriculture include aquaculture and tree farming.

The major agricultural products can be broadly grouped into foods, fibers, fuels, raw materials, pharmaceuticals and illegal drugs, and an assortment of ornamental or exotic products. In the 2000s, plants have been used to grow biofuels, biopharmaceuticals, bioplastics, and pharmaceuticals. Specific foods include cereals, vegetables, fruits, and meat. Fibers include cotton, wool, hemp, silk and flax. Raw materials include lumber and bamboo. Drugs include tobacco, alcohol, opium, cocaine,and digitalis. Other useful materials are produced by plants, such as resins. Biofuels include methane from biomass, ethanol, and biodiesel. Cut flowers, nursery plants, tropical fish and birds for the pet trade are some of the ornamental products.

In 2007, about one third of the world's workers were employed in agriculture. However, the relative significance of farming has dropped steadily since the beginning of industrialization, and in 2003 – for the first time in history – the services sector overtook agriculture as the economic sector employing the most people worldwide. Despite the fact that agriculture employs over one-third of the world's population, agricultural production accounts for less than five percent of the gross world product (an aggregate of all gross domestic products).

Etymology

The word *agriculture* is the English adaptation of Latin *agricultûra*, from *ager*, "a field", and *cultûra*, "cultivation" in the strict sense of "tillage of the soil". Thus, a literal reading of the word yields "tillage of a field / of fields".g

Overview

Agriculture has played a key role in the development of human civilization. Until the Industrial Revolution, the vast majority of the human population labored in agriculture. Development of agricultural techniques has steadily increased agricultural productivity, and the widespread diffusion of these techniques during a time period is often called an agricultural revolution. A remarkable shift in agricultural practices has occurred over the past century in response to new technologies. In particular, the Haber-Bosch method for synthesizing ammonium nitrate made the traditional practice of recycling nutrients with crop rotation and animal manure less necessary.

The percent of the human population working in agriculture has decreased over time.

Synthetic nitrogen, along with mined rock phosphate, pesticides and mechanization, have greatly increased crop yields in the early 20th century. Increased supply of grains has led to cheaper livestock as well. Further, global yield increases were experienced later in the 20th century when high-yield varieties of common staple grains such as rice, wheat, and corn (maize) were introduced as a part of the Green Revolution. The Green Revolution exported the technologies (including pesticides and synthetic nitrogen) of the developed world out to the developing world. Thomas Malthus famously predicted that the Earth would not be able to support its growing population, but technologies such as the Green Revolution have allowed the world to produce a surplus of food.

Agricultural output in 2005.

Many governments have subsidized agriculture to ensure an adequate food supply. These agricultural subsidies are often linked to the production of certain commodities such as wheat, corn (maize), rice, soybeans, and milk. These subsidies, especially when done by developed countries have been noted as protectionist, inefficient, and environmentally damaging. In the past century agriculture has been characterized by enhanced productivity, the use of synthetic fertilizers and pesticides, selective breeding, mechanization, water contamination, and farm subsidies. Proponents of organic farming such as Sir Albert Howard argued in the early 1900s that the overuse of pesticides and synthetic fertilizers damages the long-term fertility of the soil. While this feeling lay dormant for decades, as environmental awareness has increased in the 2000s there has been a movement towards sustainable agriculture by some farmers, consumers, and policymakers. In recent years there has been a backlash against perceived external environmental effects of mainstream agriculture, particularly regarding water pollution, resulting in the organic movement. One of the major forces behind this movement has been the European Union, which first certified organic food in 1991 and began reform of its Common Agricultural Policy (CAP) in 2005 to phase out commodity-linked farm subsidies, also known as decoupling. The growth of organic farming has renewed research in alternative technologies such as integrated pest management and selective breeding. Recent mainstream technological developments include genetically modified food.

As of late 2007, several factors have pushed up the price of grain used to feed poultry and dairy cows and other cattle, causing higher prices of wheat (up 58%), soybean (up 32%), and maize (up 11%) over the year. Food riots have recently taken place in many countries across the world. An epidemic of stem rust on wheat caused by race Ug99 is currently spreading across Africa and into Asia and is causing major concern. Approximately 40% of the world's agricultural land is seriously degraded. In Africa, if current trends of soil degradation continue, the continent might be able to feed just 25% of its population by 2025, according to UNU's Ghana-based Institute for Natural Resources in Africa.

History

Since its development roughly 10,000 years ago, agriculture has expanded vastly in geographical coverage and yields. Throughout this expansion, new technologies and new crops were integrated. Agricultural practices such as irrigation, crop rotation, fertilizers, and pesticides were developed long ago, but have made great strides in the past century. The history of agriculture has played a major role in human history, as agricultural progress has been a crucial factor in worldwide socio-economic change. Wealth-concentration and militaristic specializations rarely seen in hunter-gatherer cultures are commonplace in societies which practice agriculture. So, too, are arts such as epic literature and monumental architecture, as well as codified legal systems. When farmers became capable of producing food beyond the needs of their own families, others in their society were freed to devote themselves to projects other than food acquisition. Historians and anthropologists have long argued that the development of agriculture made civilization possible.

Ancient Origins

The Fertile Crescent of the Middle East, Egypt, and India were sites of the earliest planned sowing and harvesting of plants that had previously been gathered in the wild. Independent development of agriculture occurred in northern and southern China, Africa's Sahel, New Guinea and several regions of the Americas. The eight so-called Neolithic founder crops of agriculture appear: first emmer wheat and einkorn wheat, then hulled barley, peas, lentils, bitter vetch, chick peas and flax.

By 7000 BC, small-scale agriculture reached Egypt. From at least 7000 BC the Indian subcontinent saw farming of wheat and barley, as attested by archaeological excavation at Mehrgarh in Balochistan. By 6000 BC, mid-scale farming was entrenched on the banks of the Nile. About this time, agriculture was developed independently in the Far East, with rice, rather than wheat, as the primary crop. Chinese and Indonesian farmers went on to domesticate taro and beans including mung, soy and azuki. To complement these new sources of carbohydrates, highly organized net fishing of rivers, lakes and ocean shores in these areas brought in great volumes of essential protein. Collectively, these new methods of farming and fishing inaugurated a human population boom dwarfing all previous expansions, and is one that continues today.

By 5000 BC, the Sumerians had developed core agricultural techniques including large scale intensive cultivation of land, mono-cropping, organized irrigation, and use of a specialized labour force, particularly along the waterway now known as the Shatt al-Arab, from its Persian Gulf delta to the confluence of the Tigris and Euphrates. Domestication of wild aurochs and mouflon into cattle and sheep, respectively, ushered in the large-scale use of animals for food/fiber and as beasts of burden. The shepherd joined the farmer as an essential provider for sedentary and semi-nomadic societies. Maize, manioc, and arrowroot were first domesticated in the Americas as far back as 5200 BC. The potato, tomato, pepper, squash, several varieties of bean, tobacco, and several other plants were also developed in the New World, as was extensive terracing of steep hillsides in much of Andean South America. The Greeks and Romans built on techniques pioneered by the Sumerians but made few fundamentally new advances. Southern Greeks struggled with very poor soils, yet managed to become a dominant society for years. The Romans were noted for an emphasis on the cultivation of crops for trade.

A water-raising machine invented by al-Jazari (1136-1206), an Arab inventor and engineer.

Middle Ages

During the Middle Ages, Muslim farmers in North Africa and the Near East developed and disseminated agricultural technologies including irrigation systems based on hydraulic and hydrostatic principles, the use of machines such as norias, and the use of water raising machines, dams, and reservoirs. They also wrote location-specific farming manuals, and were instrumental in the wider adoption of crops including sugar cane, rice, citrus fruit, apricots, cotton, artichokes, aubergines, and saffron. Muslims also brought lemons, oranges, cotton, almonds, figs and sub-tropical crops such as bananas to Spain. The invention of a three field system of crop rotation during the Middle Ages, and the importation of the Chinese-invented moldboard plow, vastly improved agricultural efficiency.

Modern Era

After 1492, a global exchange of previously local crops and livestock breeds occurred. Key crops involved in this exchange included the tomato, maize, potato, cocoa and tobacco going from the New World to the Old, and several varieties of wheat, spices, coffee, and sugar cane going from the Old World to the New. The most important animal exportation from the Old World to the New were those of the horse and dog (dogs were already present in the pre-Columbian Americas but not in the numbers and breeds suited to farm work). Although not usually food animals, the horse (including donkeys and ponies) and dog quickly filled essential production roles on western hemisphere farms.

By the early 1800s, agricultural techniques, implements, seed stocks and cultivated plants selected and given a unique name because of its decorative or useful characteristics had so improved that yield per land unit was many times that seen in the Middle Ages. With the rapid rise of mechanization in the late 19th and 20th centuries, particularly in the form of the tractor, farming tasks could be done with a speed and on a scale previously impossible. These advances have led to efficiencies enabling certain modern farms in the United States, Argentina, Israel, Germany, and a few other nations to output volumes of high quality produce per land unit at what may be the practical limit. The Haber-Bosch method for synthesizing ammonium nitrate represented a major breakthrough and allowed crop yields to overcome previous constraints. In the past century agriculture has been characterized by enhanced productivity, the substitution of labor for synthetic fertilizers and pesticides, selective breeding, mechanization, water pollution, and farm subsidies. In recent years there has been a backlash against the external environmental effects of conventional agriculture, resulting in the organic movement.

Agricultural exploration expeditions, since the late nineteenth century, have been mounted to find new species and new agricultural practices in different areas of the world. Two early examples of expeditions include Frank N. Meyer's fruit and nut collecting trip to China and Japan from 1916-1918 and the Dorsett-Morse Oriental Agricultural Exploration Expedition to China, Japan, and Korea from 1929-1931 to collect soybean germplasm to support the rise in soybean agriculture in the

United States. In 2005, the agricultural output of China was the largest in the world, accounting for almost one-sixth world share followed by the EU, India and the USA, according to the International Monetary Fund. Economists measure the total factor productivity of agriculture and by this measure agriculture in the United States is roughly 2.6 times more productive than it was in 1948.

Crop Production Systems

Cropping systems vary among farms depending on the available resources and constraints; geography and climate of the farm; government policy; economic, social and political pressures; and the philosophy and culture of the farmer. Shifting cultivation (or slash and burn) is a system in which forests are burnt, releasing nutrients to support cultivation of annual and then perennial crops for a period of several years. Then the plot is left fallow to regrow forest, and the farmer moves to a new plot, returning after many more years (10-20). This fallow period is shortened if population density grows, requiring the input of nutrients (fertilizer or manure) and some manual pest control. Annual cultivation is the next phase of intensity in which there is no fallow period. This requires even greater nutrient and pest control inputs. Further industrialization lead to the use of monocultures, when one cultivar is planted on a large acreage. Due to the low biodiversity, nutrient use is uniform, and pests tend to build up, necessitating the greater use of pesticides and fertilizers. Multiple cropping, in which several crops are grown sequentially in one year, and intercropping, when several crops are grown at the same time are other kinds of annual cropping systems known as polycultures.

In tropical environments, all of these cropping systems are practiced. In subtropical and arid environments, the timing and extent of agriculture may be limited by rainfall, either not allowing multiple annual crops in a year, or requiring irrigation. In all of these environments perennial crops are grown (coffee, chocolate) and systems are practiced such as agroforestry. In temperate environments, where ecosystems were predominantly grassland or prairie, highly productive annual cropping is the dominant farming system. The last century has seen the intensification, concentration and specialization of agriculture, relying upon new technologies of agricultural chemicals (fertilizers and pesticides), mechanization, and plant breeding (hybrids and GMO's). In the past few decades, a move towards sustainability in agriculture has also developed, integrating ideas of socio-economic justice and conservation of resources and the environment within a farming system. This has led to the development of many responses to the conventional agriculture approach, including organic agriculture, urban agriculture, community supported agriculture, ecological or biological agriculture, integrated farming, and holistic management.

Crop Statistics

Important categories of crops include grains and pseudograins, pulses (legumes), forage, and fruits and vegetables. Specific crops are cultivated in distinct growing regions throughout the world. In millions of metric tons, based on FAO estimates.

Top agricultural products, by crop types(million metric tons) 2004 data

Top agricultural products, by crop types (million metric tons) 2004 data

Cereals	2,263
Vegetables and melons	866
Roots and Tubers	715
Milk	619
Fruit	503
Meat	259
Oilcrops	133
Fish (2001 estimate)	130
Eggs	63
Pulses	60
Vegetable Fiber	30

Source: Food and Agriculture Organization (FAO)[30]

Top agricultural products, by individual crops (million metric tons) 2004 data

Sugar Cane	1,324
Maize	721
Wheat	627
Rice	605
Potatoes	328
Sugar Beet	249
Soybean	204
Oil Palm Fruit	162
Barley	154
Tomato	120

Source: Food and Agriculture Organization (FAO)[30]

Livestock Production Systems

Animals, including horses, mules, oxen, camels, llamas, alpacas, and dogs, are often used to help cultivate fields, harvest crops, wrangle other animals, and transport farm products to buyers. Animal husbandry not only refers to the breeding and raising of animals for meat or to harvest animal products (like milk, eggs, or wool) on a continual basis, but also to the breeding and care of species for work and companionship. Livestock production systems can be defined based on feed source, as grassland-based, mixed, and landless. Grassland based livestock production relies upon plant material such as shrubland, rangeland, and pastures for feeding ruminant animals. Outside nutrient inputs may be used, however manure is returned directly to the grassland as a major nutrient source.

This system is particularly important in areas where crop production is not feasible due to climate or soil, representing 30-40 million pastoralists. Mixed production systems use grassland, fodder crops and grain feed crops as feed for ruminant and monogastic (one stomach; mainly chickens and pigs) livestock. Manure is typically recycled in mixed systems as a fertilizer for crops. Approximately 68% of all agricultural land is permanent pastures used in the production of livestock.

Landless systems rely upon feed from outside the farm, representing the de-linking of crop and livestock production found more prevalently OECD member countries. In the U.S., 70% of the grain grown is fed to animals on feedlots. Synthetic fertilizers are more heavily relied upon for crop production and manure utilization becomes a challenge as well as a source for pollution.

Production Practices

Tillage is the practice of plowing soil to prepare for planting or for nutrient incorporation or for pest control. Tillage varies in intensity from conventional to no-till. It may improve productivity by warming the soil, incorporating fertilizer and controlling weeds, but also renders soil more prone to erosion, triggers the decomposition of organic matter releasing CO_2, and reduces the abundance and diversity of soil organisms.

Pest control includes the management of weeds, insects/ mites, and diseases. Chemical (pesticides), biological (biocontrol), mechanical (tillage), and cultural practices are used. Cultural practices include crop rotation, culling, cover crops, intercropping, compost, avoidance, and resistance. Integrated pest management attempts to use all of these methods to keep pest populations below the number which would cause economic loss, and recommends pesticides as a last resort.

Nutrient management includes both the source of nutrient inputs for crop and livestock production, and the method of utilization of manure produced by livestock. Nutrient inputs can be chemical inorganic fertilizers, manure, green manure, compost and mined minerals. Crop nutrient use may also be managed using cultural techniques such as crop rotation or a fallow period. Manure is utilized either by holding livestock where the feed crop is growing such as in Managed intensive rotational grazing, or by spreading either dry or liquid formulations of manure on cropland or pastures.

Water management is where rainfall is insufficient or variable, which occurs to some degree in most regions of the world. Some farmers use irrigation to supplement rainfall. In other areas such as the Great Plains in the U.S., farmers use a fallow year to conserve soil moisture to use for growing a crop in the following year. Agriculture represents 70% of freshwater use worldwide.

Processing, Distribution, and Marketing

In the United States, food costs attributed to processing, distribution, and marketing have risen while the costs attributed to farming have declined. From 1960 to 1980 the farm share was around 40%, but by 1990 it had declined to 30% and by 1998, 22.2%. Market concentration has increased in the sector as well, with the top 20 food manufacturers accounting for half the food-processing value in 1995, over double that produced in 1954. As of 2000 the top 6 supermarkets had 50% of sales compared to 32% in 1992. Although the total effect of the increased market concentration is likely increased efficiency, the changes redistribute economic surplus from producers (farmers) and consumers, and may have negative implications for rural communities.

Crop Alteration and Biotechnology

Crop alteration has been practiced by humankind for thousands of years, since the beginning of civilization. Altering crops through breeding practices changes the genetic make-up of a plant to develop crops with more beneficial characteristics for humans, for example, larger fruits or seeds, drought-tolerance, or resistance to pests. Significant advances in plant breeding ensued after the work of geneticist Gregor Mendel. His work on dominant and recessive alleles gave plant breeders a better understanding of genetics and brought great insights to the techniques utilized by plant breeders. Crop breeding includes techniques such as plant selection with desirable traits, self-pollination and cross-pollination, and molecular techniques that genetically modify the organism. Domestication of plants has, over the centuries increased yield, improved disease resistance and drought tolerance, eased harvest and improved the taste and nutritional value of crop plants. Careful selection and breeding have had enormous effects on the characteristics of crop plants. Plant selection and breeding in the 1920s and 1930s improved pasture (grasses and clover) in New Zealand.

Extensive X-ray an ultraviolet induced mutagenesis efforts (i.e. primitive genetic engineering) during the 1950s produced the modern commercial varieties of grains such as wheat, corn (maize) and barley..

The green revolution popularized the use of conventional hybridization to increase yield many folds by creating "high-yielding varieties". For example, average yields of corn (maize) in the USA have increased from around 2.5 tons per hectare (t/ha) (40 bushels per acre) in 1900 to about 9.4 t/ha (150 bushels per acre) in 2001. Similarly, worldwide average wheat yields have increased from less than 1 t/ha in 1900 to more than 2.5 t/ha in 1990. South American average wheat yields are around 2 t/ha, African under 1 t/ha, Egypt and Arabia up to 3.5 to 4 t/ha with irrigation. In contrast, the average wheat yield in countries such as France is over 8 t/ha. Variations in yields are due mainly to variation in climate, genetics, and the level of intensive farming techniques (use of fertilizers, chemical pest control, growth control to avoid lodging))..

Genetic Engineering

Genetically Modified Organisms (GMO) are organisms whose genetic material has been altered by genetic engineering techniques generally known as recombinant DNA technology. Genetic engineering has expanded the genes available to breeders to utilize in creating desired germlines for new crops. After mechanical tomato-harvesters were developed in the early 1960s, agricultural scientists genetically modified tomatoes to be more resistant to mechanical handling. More recently, genetic engineering is being employed in various parts of the world, to create crops with other beneficial traits.

Herbicide-tolerant GMO Crops

Roundup-Ready seed has a herbicide resistance gene implanted into its genome that allows the plants to tolerate exposure to glyphosate. Roundup is a trade name for a glyphosate based product, which is a systemic, non-selective herbicide used to kill weeds. Roundup-Ready seeds allow the farmer to grow a crop that can be sprayed with glyphosate to controle weeds without harming the resistant crop. Herbicide-tolerant crops are used by farmers worldwide. Today, 92% of soybean acreage in the US is planted with genetically-modified herbicide-tolerant plants. With the increasing use of herbicide-tolerant crops, comes an increase in the use of glyphosate based herbicide sprays. In some areas glyphosate resistant weeds have developed, causing farmers to switch to other herbicides. Some studies also link widespread glyphosate usage to iron deficiencies in some crops, which is both a crop production and a nutritional quality concern, with potential economic and health implications.

Insect-Resistant GMO Crops

Other GMO crops utilized by growers include insect-resistant crops, which have a gene from the soil bacterium *Bacillus thuringiensis* (Bt) which produces a toxin specific to insects; insect-resistant crops protect plants from damage by insects, one such crop is Starlink. Another is Bt cotton, which accounts for 63% of US cotton acreage

Some believe that similar or better pest-resistance traits can be acquired through traditional breeding practices, and resistance to various pests can be gained through hybridization or cross-pollination with wild species. In some cases, wild species are the primary source of resistance traits; some Tomato cultivars that have gained resistance to at least nineteen diseases, did so, through crossing with wild populations of tomatoes.

Costs and Benefits of GMOs

Genetic engineers may someday develop transgenic plants which would allow for irrigation, drainage, conservation, sanitary engineering, and maintaining or increasing yields while requiring fewer fossil fuel derived inputs than conventional crops.[22] Such developments would be particularly important in areas which are normally arid and rely upon constant irrigation, and on large scale farms. However, genetic engineering of plants has proven to be controversial. Many issues surrounding food security and environmental impacts have risen regarding GMO practices. For example, GMOs are questioned by some ecologists and economists concerned with

GMO practices such as terminator seeds,, which is a genetic modification that creates sterile seeds. Terminator seeds are currently under strong international opposition and face continual efforts of global bans. Another controversial issue is the patent protection given to companies that develop new types of seed using genetic engineering. Since companies have intellectual ownership of their seeds, they have the power to dictate terms and conditions of their patented product. Currently, ten seed companies control over two-thirds of the global seed sales. Vandana Shiva argues that these companies are guilty of biopiracy by patenting life and exploiting organisms for profit Farmers using patented seed are restricted from saving seed for subsequent plantings, which forces farmers to buy new seed every year. Since seed saving is a traditional practice for many farmers in both developing and developed countries, GMO seeds legally bind farmers to change their seed saving practices to buying new seed every year.

Locally adapted seeds are an essential hertitage that has the potential to be lost with current hybridized crops and GMOs. Locally adapted seeds, also called land races or crop ecotypes, are important because they have adapted over time to the specific microclimates, soils, other environmental conditions, field designs, and ethnic preference indigenous to the exact area of cultivation Introducing GMOs and hybridized commercial seed to an area brings the risk of cross-pollination with local land races Therefore, GMOs pose a threat to the sustainability of land races and the ethnic heritage of cultures. Once seed contains transgenic material, it becomes subject to the conditions of the seed company that owns the patent of the transgenic material. There is also concern that GMOs will cross-pollinate with wild species and permanently alter native populations' genetic integrity; there are already identified populations of wild plants with transgenic genes. GMO gene flow to related weed species is a concern, as well as cross-pollination with non-transgenic crops. Since many GMO crops are harvested for their seed, such as rapeseed, seed spillage in is problematic for volunteer plants in rotated fields, as well as seed-spillage during transportation.

Food Safety and Labeling

Food security issues also coincide with food safety and food labeling concerns. Currently a global treaty, the BioSafety Protocol, regulates the trade of GMOs. The EU currently requires all GMO foods to be labeled, whereas the US does not require transparent labeling of GMO foods. Since there are still questions regarding the safety and risks associated with GMO foods, some believe the public should have the freedom to choose and know what they are eating and require all GMO products to be labeled.

ENVIRONMENTAL IMPACT

External Costs

Agriculture imposes external costs upon society through pesticides, nutrient runoff, excessive water usage, and assorted other problems. A 2000 assessment of agriculture in the UK determined total external costs costs for 1996 of 2343 million British pounds or 208 pounds per hectare. A 2005 analysis of these costs in the USA concluded that cropland imposes approximately 5 to 16 billion dollars ($30 to $96 per hectare), while livestock production imposes 714 million dollars. Both studies concluded that more should be done to internalize external costs, and neither included subsidies in their analysis, but noted that subsidies also influence the cost of agriculture to society. Both focused on purely fiscal impacts. The 2000 review included reported pesticide poisonings but did not include speculative chronic effects of pesticides, and the 2004 review relied on a 1992 estimate of the total impact of pesticides.

Land Transformation and Degradation

Land transformation, the use of land to yield goods and services, is the most substantial way humans alter the Earth's ecosystems, and is considered the driving force in the loss of biodiversity. Estimates of the amount of land transformed by humans vary from 39–50%. Land degradation, the long-term decline in ecosystem function and productivity, is estimated to be occurring on 24% of land worldwide, with cropland overrepresented. The UN-FAO report cites land management as the driving factor behind degradation and reports that 1.5 billion people rely upon the degrading land. Degradation can be deforestation, desertification, soil erosion, mineral depletion, or chemical degradation (acidification and salinization).

Eutrophication

Eutrophication, excessive nutrients in aquatic ecosystems resulting in algal blooms and anoxia, leads to fish kills, loss of biodiversity, and renders water unfit for drinking and other industrial uses. Excessive fertilization and manure application to cropland, as well as high livestock stocking densities cause nutrient (mainly nitrogen and phosphorus) runoff and leaching from agricultural land. These nutrients are major nonpoint pollutants contributing to eutrophication of aquatic ecosystems.

Pesticides

Pesticide use has increased since 1950 to 2.5 million tons annually worldwide, yet crop loss due to pests has remained relatively constant. The World Health Organization estimated in 1992 that 3 million pesticide poisonings occur annually, causing 220,000 deaths. Pesticides select for pesticide resistance in the pest population, leading to a condition termed the 'pesticide treadmill' in which pest resistance warrants the development of a new pesticide. An alternative argument is that the way to 'save the environment' and prevent famine is by using pesticdes and intensive high yield farming, a view exemplified by a quote heading the Center for Global Food Issues website: 'Growing more per acre leaves more land for nature'. However critics argue that a tradeoff between the environment and a need for food is not inevitable, and that pesticides simply replace good agronomic practices such as crop rotation.

A senior UN official and co-author of a UN report detailing this problem, Henning Steinfeld, said "Livestock are one of the most significant contributors to today's most serious environmental problems." Livestock production occupies 70% of all land used for agriculture, or 30% of the land surface of the planet. It is one of the largest sources of greenhouse gases, responsible for 18% of the world's greenhouse gas emissions as measured in CO_2 equivalents. By comparison, all transportation emits 13.5% of the CO_2. It produces 65% of human-related nitrous oxide (which has 296 times the global warming potential of CO_2) and 37% of all human-induced methane (which is 23 times as warming as CO_2). It also generates 64% of the ammonia, which contributes to acid rain and acidification of ecosystems. Livestock expansion is cited as a key factor driving deforestation, in the Amazon basin 70% of previously forested area is now occupied by pastures and the remainder used for feedcrops. Through deforestation and land degradation, livestock is also driving reductions in biodiversity.

Climate Change

Climate change has the potential to affect agriculture through changes in temperature and moisture regimes. Agriculture can both mitigate or worsen global warming. Some of the increase in CO_2 in the atmosphere comes from the decomposition of organic matter in the soil, and much of the methane emitted into the atmosphere is due to the decomposition of organic matter in wet soils such as rice paddies. Further, wet or anaerobic soils also lose nitrogen through denitrification, releasing the greenhouse gas nitric oxide. Changes in management can reduce the release of these greenhouse gases, and soil can further be used to sequester some of the CO_2 in the atmosphere.

Distortions in Modern Global Agriculture

Differences in economic development, population density and culture mean that the farmers of the world operate under very different conditions.

A US cotton farmer may receive US$230 government subsidies per acre planted (as in 2003), farmers in Mali and other third world countries do without. When prices decline, the heavily subsidised US farmer is not forced to reduce his output, hence making it difficult for cotton prices to rebound, his Mali counterpart may go broke in the meantime.

A livestock farmer in South Korea can calculate with a (highly subsedised) salesprice of US$1300 for a calf produced.

A South American Mercosur country rancher calculates with a calf's salesprice of US$120-200 (both 2008 figures). With the former, scarcety and high cost of land is compensated with public subsidies, the latter compensates absence of subsedies with economics of scale and low cost of land.

In PR China a rural household's productive asset may be one hectare of farmland. In Brazil, Paraguay and other countries where local legislature allows such purchases, international investors buy thousands of hectares of farmland or raw land at prices of a few hundred US$ per hectare.

Agriculture and Petroleum

Since the 1940s, agriculture has dramatically increased its productivity, due largely to the use of petrochemical derived pesticides, fertilizers, and increased mechanization (the so-called Green Revolution). Between 1950 and 1984, as the Green Revolution transformed agriculture around the globe, world grain production increased by 250%. This has allowed world population to grow more than double over the last 50 years. However, every energy unit delivered in food grown using modern techniques requires over ten energy units to produce and deliver, although this statistic is contested by proponents of petroleum-based agriculture. The vast majority of this energy input comes from fossil fuel sources. Because of modern agriculture's current heavy reliance on petrochemicals and mechanization, there are warnings that the ever decreasing supply of oil (the dramatic nature of which is known as peak oil) will inflict major damage on the modern industrial agriculture system, and could cause large food shortages.

Modern or industrialized agriculture is dependent on petroleum in two fundamental ways: 1) cultivation—to get the crop from seed to harvest and 2) transport—to get the harvest from the farm to the consumer's refrigerator. It takes approximately 400 gallons of oil a year per citizen to fuel the tractors, combines and other equipment used on farms for cultivation or 17 percent of the nation's total energy use. Oil and natural gas are also the building blocks of the fertilizers, pesticides and herbicides used on farms. Petroleum is also providing the energy required to process food before it reaches the market. It takes the energy equivalent of a half-gallon of gasoline to produce a two-pound bag of breakfast cereal. And that still does not count the energy needed to transport that cereal to market; it is the transport of processed foods and crops that consumes the most oil. The kiwi from New Zealand, the asparagus from Argentina, the melons and broccoli from Guatemala, the organic lettuce from California-most food items on the consumer's plate travel average of 1,500 miles just to get there.

Oil shortages could interrupt this food supply. The consumer's growing awareness of this vulnerability is one of several factors fueling current interest in organic agriculture and other sustainable farming methods. Some farmers using modern organic-farming methods have reported yields as high as those available from conventional farming (but without the use of fossil-fuel-intensive artificial fertilizers or pesticides. However, the reconditioning of soil to restore nutrients lost during the use of monoculture agriculture techniques made possible by petroleum-based technology will take time.

The dependence on oil and vulnerability of the U.S. food supply has also led to the creation of a conscious consumption movement in which consumers count the "food miles" a food product has traveled. The Leopold Center for Sustainable Agriculture defines a food mile as: "...the distance food travels from where it is grown or raised to where it is ultimately purchased by the consumer or end-user." In a comparison of locally-grown food and long-distance food, researchers at the Leopold Center found that local food traveled an average of 44.6 miles to reach its destination compared with 1,546 miles for conventionally-grown and shipped food.

Consumers in the new local food movement who count food miles call themselves "locavores" LINK; they advocate a return to a locally-based food system where food comes from as close as possible, whether or not it is organic. Locavores argue that an organically-grown lettuce from California that is shipped to New York is still an unsustainable food source because of dependence on fossil fuels to ship it. In addition to the "locavore" movement, concern over dependence on oil-based agriculture has also dramatically increased interest in home and community gardening.LINK

Farmers have also begun raising crops such as corn (maize) for non-food use in an effort to help mitigate peak oil. This has contributed to a 60% rise in wheat prices recently, and has been indicated as a possible precursor to "serious social unrest in developing countries." Such situations would be exacerbated in the event of future rises in food and fuel costs, factors which have already impacted the ability of charitable donors to send food aid to starving populations.

One example of the chain reactions which could possibly be caused by peak oil issues involves the problems caused by farmers raising crops such as corn (maize) for non-food use in an effort to help mitigate peak oil. This has already lowered food production. This food vs fuel issue will be exacerbated as demand for ethanol fuel rises. Rising food and fuel costs has already limited the abilities of some charitable donors to send food aid to starving populations. In the UN, some warn that the recent 60% rise in wheat prices could cause "serious social unrest in developing countries." In 2007, higher incentives for farmers to grow non-food biofuel crops combined with other factors (such as over-development of former farm lands, rising transportation costs, climate change, growing consumer demand in China and India, and population growth) to cause food shortages in Asia, the Middle East, Africa, and Mexico, as well as rising food prices around the globe. As of December 2007, 37 countries faced food crises, and 20 had imposed some sort of food-price controls. Some of these shortages resulted in food riots and even deadly stampedes.

Another major petroleum issue in agriculture is the effect of petroleum supplies will have on fertilizer production. By far the biggest fossil fuel input to agriculture is the use of natural gas as a hydrogen source for the Haber-Bosch fertilizer-creation process. Natural gas is used because it is the cheapest currently available source of hydrogen. When oil production becomes so scarce that natural gas is used as a partial stopgap replacement, and hydrogen use in transportation increases, natural gas will become much more expensive. If the Haber Process is unable to be commercialized using renewable energy (such as by electrolysis) or if other sources of hydrogen are not available to replace the Haber Process, in amounts sufficient to supply transportation and agricultural needs, this major source of fertilizer would either become extremely expensive or unavailable. This would either cause food shortages or dramatic rises in food prices.

Mitigation of Effects of Petroleum Shortages

One effect oil shortages could have on agriculture is a full return to organic agriculture. In light of peak oil concerns, organic methods are much more sustainable than contemporary practices because they use no petroleum-based pesticides, herbicides, or fertilizers. Some farmers using modern organic-farming methods have reported yields as high as those available from conventional farming. Organic farming may however be more labor-intensive and would require a shift of work force from urban to rural areas.

It has been suggested that rural communities might obtain fuel from the biochar and synfuel process, which uses agricultural *waste* to provide charcoal fertilizer, some fuel *and* food, instead of the normal food vs fuel debate. As the synfuel would be used on site, the process would be more efficient and may just provide enough fuel for a new organic-agriculture fusion.

It has been suggested that some transgenic plants may some day be developed which would allow for maintaining or increasing yields while requiring fewer fossil fuel derived inputs than conventional crops. The possibility of success of these programs is questioned by ecologists and economists concerned with unsustainable GMO practices such as terminator seeds, and a January 2008 report shows that GMO practices "fail to deliver environmental, social and economic benefits." While there has been some research on sustainability using GMO crops, at least one hyped and prominent multi-year attempt by Monsanto has been unsuccessful, though during the same period traditional breeding techniques yielded a more sustainable variety of the same crop. Additionally, a survey by the bio-tech industry of subsistence farmers in Africa to discover what GMO research would most benefit sustainable agriculture only identified non-transgenic issues as areas needing to be addressed. Nonetheless, some governments in Africa continue to view investments in new transgenic technologies as an essential component of efforts to improve sustainability.

Policy

Agricultural policy focuses on the goals and methods of agricultural production. At the policy level, common goals of agriculture include:

- Conservation
- Economic stability
- Environmental impact
- Food quality: Ensuring that the food supply is of a consistent and known quality.
- Food safety: Ensuring that the food supply is free of contamination.
- Food security: Ensuring that the food supply meets the population's needs.
- Poverty Reduction

AGRICULTURE SAFETY AND HEALTH

United States

Agriculture ranks among the most hazardous industries. Farmers are at high risk for fatal and nonfatal injuries, work-related lung diseases, noise-induced hearing loss, skin diseases, and certain cancers associated with chemical use and prolonged sun exposure. Farming is one of the few industries in which the families (who often share the work and live on the premises) are also at risk for injuries, illness, and death. In an average year, 516 workers die doing farm work in the U.S. (1992-2005). Of these deaths, 101 are caused by tractor overturns. Every day, about 243 agricultural workers suffer lost-work-time injuries, and about 5% of these result in permanent impairment.

Agriculture is the most dangerous industry for young workers, accounting for 42% of all work-related fatalities of young workers in the U.S. between 1992 and 2000. Unlike other industries, half the young victims in agriculture were under age 15. For young agricultural workers aged 15–17, the risk of fatal injury is four times the risk for young workers in other workplaces Agricultural work exposes young workers to safety hazards such as machinery, confined spaces, work at elevations, and work around livestock.

An estimated 1.26 million children and adolescents under 20 years of age resided on farms in 2004, with about 699,000 of these youth performing work on the farms. In addition to the youth who live on farms, an additional 337,000 children and adolescents were hired to work on U.S. farms in 2004. On average, 103 children are killed annually on farms (1990-1996). Approximately 40 percent of these deaths were work-related. In 2004, an estimated 27,600 children and adolescents were injured on farms; 8,100 of these injuries were due to farm work.

HISTORY OF AGRICULTURE

Agriculture was developed at least 10,000 years ago, and it has undergone significant developments since the time of the earliest cultivation. Evidence points to the Fertile Crescent of the Middle East as the site of the earliest planned sowing and harvesting of plants that had previously been gathered in the wild. Independent development of agriculture is also believed to have occurred in northern and southern China, Africa's Sahel, New Guinea and several regions of the Americas. Agricultural practices such as irrigation, crop rotation, fertilizers, and pesticides were developed long ago but have made great strides in the past century. The Haber-Bosch method for synthesizing ammonium nitrate represented a major breakthrough and allowed crop yields to overcome previous constraints. In the past century, agriculture has been characterized by enhanced productivity, the substitution of human labor for synthetic fertilizers and pesticides, selective breeding, and mechanization. The recent history of agriculture has been closely tied with a range of political issues including water pollution, biofuels, Genetically modified organisms, tariffs, and farm subsidies. In recent years, there has been a backlash against the external environmental effects of mechanized agriculture, and increasing support for the organic movement and Sustainable agriculture.

Timeline

Scholars have proposed a number of theories to explain the historical development of farming. Most likely, there was a gradual transition from hunter-gatherer to agricultural societies after a lengthy period during which some crops were deliberately planted and other foods were gathered in the wild. Although localised climate change is the favoured explanation for the origins of agriculture in the Levant, the fact that farming was 'invented' at least three times elsewhere, suggests that social reasons may have been instrumental. When major climate change took place after the last ice age (c.11,000 BC), much of the earth became subject to long dry seasons. These conditions favoured annual plants which die off in the long dry season, leaving a dormant seed or tuber. These plants tended to put more energy into producing seeds than into woody growth. An abundance of readily storable wild grains and pulses enabled hunter-gatherers in some areas to form the first settled villages at this time.

The Oasis Theory was proposed by Raphael Pumpelly in 1908, and popularized by Vere Gordon Childe who summarized the theory in his book *Man Makes Himself* This theory maintains that as the climate got drier, communities contracted to oases where they were forced into close association with animals which were then domesticated together with planting of seeds. The theory has little support from contemporary scholars, as the climatex data for the time does not support the theory.

The Hilly Flanks hypothesis, proposed by Robert Braidwood in 1948, suggests that agriculture began in the hilly flanks of the Taurus and Zagros mountains, and that it developed from intensive focused grain gathering in the region.

The Feasting model by Bryan Hayden suggests that agriculture was driven by ostentatious displays of power, such as throwing feasts to exert dominance. This required assembling large quantities of food which drove agricultural technology.

The Demographic theories were proposed by Carl Sauer and adapted by Lewis Binford and Kent Flannery. They describe an increasingly sedentary population, expanding up to the carrying capacity of the local environment, and requiring more food than can be gathered. Various social and economic factors help drive the need for food. The evolutionary/intentionality theory, advanced by scholars including David Rindos, is the idea that agriculture is an co-evolutionary adaptation of plants and humans. Starting with domestication by protection of wild plants, followed specialisation of location and then domestication.

Neolithic Era

Identifying an exact origin of agriculture remains problematic because the transition from hunter-gatherer societies began thousands of years before the invention of writing. Nonetheless, archaeobotanists/paleoethnobotanists have traced the selection and cultivation of specific food plant characteristics, such as a semi-tough rachis and larger seeds, to just after the Younger Dryas (about 9,500 BC) in the early Holocene in the Levant region of the Fertile Crescent. There is earlier evidence for use of wild cereals: anthropological and archaeological evidence from sites across Southwest Asia and North Africa indicate use of wild grain (e.g., from the ca. 20,000 BC site of Ohalo II in Israel, many Natufian sites in the Levant and from sites along the Nile in the 10th millennium BC). There is even evidence of planned cultivation and trait selection: grains of rye with domestic traits have been recovered from Epi-Palaeolithic (10,000+ BC) contexts at Abu Hureyra in Syria, but this appears to be a localised phenomenon resulting from cultivation of stands of wild rye, rather than a definitive step towards domestication. It isn't until after 9,500 BC that the eight so-called founder crops of agriculture appear: first emmer and einkorn wheat, then hulled barley, peas, lentils, bitter vetch, chick peas and flax. These eight crops occur more or less simultaneously on PPNB sites in the Levant, although the consensus is that wheat was the first to be sown and harvested on a significant scale.

By 7000 BC, sowing and harvesting reached Mesopotamia and there, in the fertile soil just north of the Persian Gulf, Sumerians systematized it and scaled it up. By 6000 BC farming was entrenched on the banks of the Nile River. About this time, agriculture was developed independently in the Far East, probably in China, with rice rather than wheat as the primary crop. Maize was first domesticated, probably from teosinte, in the Americas around 3000-2700 BC, though there is some archaeological evidence of a much older development. The potato, the tomato, the pepper, squash, several varieties of bean, and several other plants were also developed in the New World, as was quite extensive terracing of steep hillsides in much of Andean South America. Agriculture was also independently developed on the island of New Guinea

In Europe, there is evidence of emmer and einkorn wheat, barley, sheep, goats and pigs that suggest a food producing economy in Greece and the Aegean by 7000 BC. Archaeological evidence from various sites on the Iberian peninsula suggest the domestication of plants and animals between 6000 and 4500 BC. Céide Fields in Ireland, consisting of extensive tracts of land enclosed by stone walls, date to 5500 BC and are the oldest known field systems in the world. The horse was domesticated in Ukraine around 4000 BC.

In China, rice and millet were domesticated by 8000 BC, followed by the beans mung, soy and azuki. In the Sahel region of Africa local rice and sorghum were domestic by 5000 BC. Local crops were domesticated independently in West Africa and possibly in New Guinea and Ethiopia. Evidence of the presence of wheat and some legumes in the 6th millennium BC have been found in the Indus Valley. Oranges were cultivated in the same millennium. The crops grown in the valley around 4000 BC were typically wheat, peas, sesame seed, barley, dates and mangoes. By 3500 BC cotton growing and cotton textiles were quite advanced in the valley. By 3000 BC farming of rice had started. Other monsoon crops of importance of the time was cane sugar. By 2500 BC, rice was an important component of the staple diet in Mohenjodaro near the Arabian Sea. By this time the Indians had large cities with well-stocked granaries. Three regions of the Americas independently domesticated corn, squashes, potato and sunflowers.

Bronze Age

By the Bronze Age, wild food contributed a nutritionally insignificant component to the usual diet. If the operative definition of *agriculture* includes large scale intensive cultivation of land, mono-cropping, organized irrigation, and use of a specialized labour force, the title "inventors of agriculture" would fall to the Sumerians, starting ca. 5,500 BC. Intensive farming allows a much greater density of population than can be supported by hunting and gathering, and allows for the accumulation of excess product for off-season use, or to sell/ barter. The ability of farmers to feed large numbers of people whose activities have nothing to do with agriculture was the crucial factor in the rise of standing armies. Sumerian agriculture supported a substantial territorial expansion which along with internecine conflict between cities, made them the first empire builders. Not long after, the Egyptians, powered by farming in the fertile Nile valley, achieved a population density from which enough warriors could be drawn for a territorial expansion more than tripling the Sumerian empire in area.

In Sumer, barley was the primary crop; wheat, flax, dates, apples, plums, and grapes were grown as well. Mesopotamia was blessed with flooding from the Tigris and Euphrates rivers, but floods came in late spring or early summer from snow melting from the Anatolian mountains. This, along with salt deposits in the soil, made farming in Mesopotamia difficult. Sheep and goats were also domesticated, kept mainly for meat and milk, butter and cheese being made from the latter. Ur, a large town that covered about 50 acres (20 hectares), had 10,000 animals kept in sheepfolds and stables and 3,000 slaughtered every year. The city's population of 6,000 included a labour force of 2,500, cultivated 3,000 acres (12 km^2) of land. The labour force contained storehouse recorders, work foremen, overseers, and harvest supervisors to supplement labourers.

Agricultural produce was given to temple personnel, important people in the community, and small farmers.

The land was plowed by teams of oxen pulling light unwheeled plows and grain was harvested with sickles in the spring. Wagons had solid wheels covered by leather tires kept in position by copper nails and were drawn by oxen and the Syrian onager (now extinct). Animals were harnessed by collars, yokes, and headstalls. They were controlled by reins, and a ring through the nose or upper lip and a strap under the jaw. As many as four animals could pull a wagon at one time. The horse was domesticated in Ukraine around 4000 BC, and was in use by the Sumerians around 2000 BC.

Roman Era

Roman agriculture built off techniques pioneered by the Sumerians, transmitted to them by subsequent cultures, with a specific emphasis on the cultivation of crops for trade and export. Romans laid the groundwork for the manorial economic system, involving serfdom, which flourished in the Middle Ages. The farm sizes in Rome can be divided into three categories. Small farms were from 18-88 iugera (one iugera is equal to about 0.65 acre). Medium-sized farms were from 80-500 iugera (singular iugerum). Large estates (called latifundia) were over 500 iugera.

The Romans had four systems of farm management: direct work by owner and his family; slaves doing work under supervision of slave managers; tenant farming or sharecropping in which the owner and a tenant divide up a farm's produce; and situations in which a farm was leased to a tenant. There was a great deal of commerce between the provinces of the empire, all the regions of the empire became interdependent with one another, some provinces specialized in the production of grain, others in wine and others in olive oil, depending on the soil type.

Middle Ages

During the Middle Ages, Muslim farmers in North Africa and the Near East developed and disseminated agricultural technologies including irrigation systems based on hydraulic and hydrostatic principles, the use of machines such as norias, and the use of water raising machines, dams, and reservoirs. They also wrote location-specific farming manuals, and were instrumental in the wider adoption of crops including sugar cane, rice, citrus fruit, apricots, cotton, artichokes, aubergines, and saffron. Muslims also brought lemons, oranges, cotton, almonds, figs and sub-tropical crops such as bananas to Spain.

The invention of a three field system of crop rotation during the Middle Ages, and the importation of the Chinese-invented moldboard plow, vastly improved agricultural efficiency. After 1492 the world's agricultural patterns were shuffled in the widespread exchange of plants and animals known as the Columbian Exchange. Crops and animals that were previously only known in the Old World were now transplanted to the New and vice versa. Perhaps most notably, the tomato became a favorite in European cuisine, and maize and potatoes were widely adopted. Other transplanted crops include pineapple, cocoa, and tobacco. In the other direction, several wheat strains quickly took to western hemisphere soils and became a dietary staple even for native North, Central and South Americans. Agriculture was a key element in the Atlantic slave trade, Triangular trade, and the expansion by European powers into the Americas. In the expanding Plantation economy, large plantations producing crops including sugar, cotton, and indigo, were heavily dependent upon slave labor.

REGIONAL HISTORIES

China

The unique tradition of Chinese agriculture has been traced to the pre-historic Xianrendong Relics and Diaotonghuan Relics (c. 12 0000 BC-7500 BC). Chinese historical and governmental records of the Warring States (481 BC-221 BC), Qin Dynasty (221 BC-207 BC), and Han Dynasty (202 BC-220 AD) eras allude to the use of complex agricultural practices, such as a nationwide granary system and widespread use of sericulture. However, the oldest extant Chinese book on agriculture is the *Chimin Yaoshu* of 535 AD, written by Jia Sixia. Although much of the literature of the time was elaborate, flowery, and allusive, Jia's writing style was very straightforward and lucid, a literary approach to agriculture that later Chinese agronomists after Jia would follow, such as Wang Zhen and his groundbreaking *Nong Shu* of 1313 AD. Jia's book was also incredibly long, with over one hundred thousand written Chinese characters, and quoted 160 other Chinese books that were written previously (but no longer survive). The contents of Jia's 6th century book include sections on land preparation, seeding, cultivation, orchard management, forestry, and animal husbandry. The book also includes peripherally related content covering trade and culinary uses for crops.

For agricultural purposes, the Chinese had innovated the hydraulic-powered trip hammer by the 1st century BC. Although it found other purposes, its main function to pound, decorticate, and polish grain that otherwise would have been done manually. The Chinese also innovated the square-pallet chain pump by the 1st century AD, powered by a waterwheel or an oxen pulling a on a system of mechanical wheels. Although the chain pump found use in public works of providing water for urban and palatial pipe systems, it was used largely to lift water from a lower to higher elevation in filling irrigation canals and channels for farmland.

Papua

Ancient Papuans are thought to have begun practicing agriculture around 7000 BC. They began domesticating sugarcane and root crops. Pigs may also have been domesticated around this time. By 3000 BC, Papuan agriculture was characterised by water control for irrigation.

India

Wheat, barley, and jujube were domesticated in the Indian subcontinent by 9000 BCE; Domestication of sheep and goat soon followed. Barley and wheat cultivation—along with the domestication of cattle, primarily sheep and goat—continued in Mehrgarh culture by 8000-6000 BCE. This period also saw the first domestication of the elephant. Agro pastoralism in India included threshing, planting crops in rows—either of two or of six—and storing grain in granaries. By the 5th millennium BCE agricultural communities became widespread in Kashmir. Cotton was cultivated by the 5th millennium BCE-4th millennium BCE.

Archaeological evidence indicates that rice was a part of the Indian diet by 8000 BCE. The Encyclopedia Britannica—on the subject of the first certain cultivated rice—holds that: A number of cultures have evidence of early rice cultivation, including China, India, and the civilizations of Southeast Asia.

Irrigation was developed in the Indus Valley Civilization by around 4500 BCE. The size and prosperity of the Indus civilization grew as a result of this innovation, which eventually led to more planned settlements making use of drainage and sewers. Archeological evidence of an animal-drawn plough dates back to 2500 BC in the Indus Valley Civilization.

Mesoamerica

In Mesoamerica, the Aztecs were some of the most innovative farmers of the ancient world and farming provided the entire basis of their economy. The land around Lake Texcoco was fertile, but not large enough to produce the amount of food needed for the population of their expanding empire. The Aztecs developed irrigation systems, formed terraced hillsides, and fertilized their soil. However, their greatest agricultural technique was the chinampas, or artificial islands, also known as "floating gardens". These were used to make the swampy areas around the lake suitable for farming. To make chinampas, canals were dug through the marshy islands and shores, then mud was heaped on huge mats made of woven reeds. The mats were anchored by tying them to posts driven into the lake bed and then planting trees at their corners that took root and secured the artificial islands permanently. The Aztecs grew corn, squash, vegetables, and flowers on chinampas.

Incas

The Andean civilizations were predominantly agricultural societies; the Incas took advantage of the ground, conquering the adversities like the Andean area and the inclemencies of the weather. The adaptation of agricultural technologies that already were used previously, allowed the Incas to organize the production a diversity of products of the coast, mountain and jungle, so them could be able to redistribute to villages that did not have access to other regions. The technological achievements reached to agricultural level, had not been possible without the workforce that was at the disposal of the Sapa Inca, as well as the road system that was allowing to store adequately the harvested resources and to distribute them for all the territory.

Islamic World

From the 8th century, the medieval Islamic world witnessed a fundamental transformation in agriculture known as the "Muslim Agricultural Revolution" or "Arab Agricultural Revolution". Due to the trade routes established by Muslim traders across the Old World during the "Afro-Asiatic age of discovery" or "Pax Islamica", this enabled the diffusion of many crops, plants and farming techniques across the Islamic world, as well as the adaptation of crops, plants and techniques from beyond the Islamic world. These techniques included crop rotation, irrigation and pest control. Some have referred to the diffusion of numerous crops during this period as the "Globalisation of Crops",which, along with increased mechanization of agriculture, led to major changes in economy, population distribution, vegetation cover, agricultural production and income, population levels, urban growth, the distribution of the labour force, linked industries, cooking and diet, clothing, and numerous other aspects of life in the Islamic world.

Serfdom became widespread in eastern Europe in the Middle Ages. Medieval Europe owed much of its development to advances made in Islamic areas (see Islamic contributions to Medieval Europe), which flourished culturally and materially while Europe and other Roman and Byzantine administered lands entered an extended period of social and economic stagnation. As early as the ninth century, an essentially modern agricultural system became central to economic life and organization in the Arab caliphates, replacing the largely export driven Roman model. The cities of the Near East, North Africa and Moorish Spain were supported by agricultural systems which included irrigation based on knowledge of hydraulic and hydrostatic principles, some of which were continued from Roman times. In later centuries, Persian Muslims transmitted cultural elements, including advanced agricultural techniques, into Turkic lands and western India.

The Muslim agricultural revolution was based on the development of a sophisticated system of irrigation using machines such as norias, water mills, water raising machines, dams and reservoirs. With such technology they managed to greatly expand the exploitable land area. As well, they adopted a scientific approach to farming enabled them to improve farming techniques derived from the collection and collation of relevant information throughout the whole of the known world. Farming manuals were widely produced, detailing where, when, and how to plant and grow various crops. Reformed land ownership rules and labourers' rights, combining the recognition of private ownership and the rewarding of cultivators with a harvest share commensurate with their efforts also improved agricultural practices. Their counterparts in Europe struggled under a feudal system in which they were almost slaves (serfs) with little hope of improving their lot by hard work.

The introduction of new crops transforming private farming into a new global industry exported everywhere, including Europe, where farming was mostly restricted to wheat strains obtained much earlier via central Asia. Spain received what she in turn transmitted to the rest of Europe; many agricultural and fruit-growing processes, together with many new plants, fruit and vegetables. These new crops included sugar cane, rice, citrus fruit, apricots, cotton, artichokes, aubergines, and saffron. Others, previously known, were further developed. Muslims also brought to that country lemons, oranges, cotton, almonds, figs and sub-tropical crops such as bananas and sugar cane. Several were later exported from Spanish coastal areas to the Spanish colonies in the New World. Also transmitted via Muslim influence, a silk industry flourished, flax was cultivated and linen exported, and esparto grass, which grew wild in the more arid parts, was collected and turned into various articles.

Britain

Between the 16th century and the mid-19th century, Great Britain saw a massive increase in agricultural productivity and net output. New agricultural practices like enclosure, mechanization, four-field crop rotation and selective breeding enabled an unprecedented population growth, freeing up a significant percentage of the workforce, and thereby helped drive the Industrial Revolution. By the early 1800s, agricultural practices, particularly careful selection of hardy strains and cultivars, had so improved that yield per land unit was many times that seen in the Middle Ages and before.

The 18th and 19th century also saw the development of glasshouses, or greenhouses, initially for the protection and cultivation of exotic plants imported to Europe and North America from the tropics. Experiments on Plant Hybridization in the late 1800s yielded advances in the understanding of plant genetics, and subsequently, the development of hybrid crops. Increasing dependence upon monoculture crops lead to famines and food shortages, most notably the Irish Potato Famine (1845–1849). Storage silos and grain elevators appeared in the 19th century.

Industrialization

With the rapid rise of mechanization in the late 19th and 20th centuries, particularly in the form of the tractor, and later the Combine harvester, farming tasks could be done with a speed and on a scale previously impossible. These advances, joined to science-driven innovations in methods and resources, have led to efficiencies enabling certain modern farms in the United States, Argentina, Israel, Germany and a few other nations to output volumes of high quality produce per land unit at what may be the practical limit. The development of rail and highway networks and the increasing use of container shipping and refrigeration in developed nations have also been essential to the growth of mechanized agriculture, allowing for the economical long distance shipping of produce.

While chemical fertilizer and pesticide have existed since the 19th century, their use grew significantly in the early twentieth century. Until the development of chemical fertilizers, Guano was widely used as a fertilizer, though expensive. The Haber-Bosch method for synthesizing ammonium nitrate represented a major breakthrough and allowed crop yields to overcome previous constraints. It was first patented by German chemist Fritz Haber. In 1910 Carl Bosch, while working for German chemical company BASF, successfully commercialized the process and secured further patents. Norman Borlaug and other scientists began developing crops for increased yields in the 1940s in Mexico. Their work lead to the Green Revolution, which applied western advances in fertilizer and pesticide use to farms worldwide, with varying success. Other applications of scientific research since 1950 in agriculture include gene manipulation, Hydroponics, and the development of economically viable biofuels such as Ethanol.

Though the intensive fishing practices pioneered and extended in recent history generally led to increased outputs, they have also led to the destruction of farmland, most notably in the dust bowl area of the United States following World War I. As global population increases, agriculture continues to replace natural ecosystems with monoculture crops. Since the 1970s, western farmers and consumers have become increasingly aware of, and in some cases critical of, widely used intensive agriculture practices. This growing awareness has lead to increased interest in such areas of agriculture as organic farming, permaculture, Heirloom plants and biodiversity, the growth of the Slow Food movement, and an ongoing discussion surrounding the potential for sustainable agriculture.

AGRIBUSINESS

In agriculture, agribusiness is a generic term that refers to the various businesses involved in food production, including farming and contract farming, seed supply, agrichemicals, farm machinery, wholesale and distribution, processing, marketing, and retail sales. The term has two distinctly different connotations depending on context.

Within the agriculture industry, agribusiness is widely used simply as a convenient portmanteau of agriculture and business, referring to the range of activities and disciplines encompassed by modern food production. There are academic degrees in and departments of agribusiness, agribusiness trade associations, agribusiness publications, and so forth, worldwide. Here, the term is only descriptive, and is synonymous in the broadest sense with food industry. Among critics of large-scale, industrialized, vertically integrated food production, the term *agribusiness* is used as a negative, synonymous with corporate farming. As such, it is often contrasted with family farm. Some negative connotation is also derived from the negative associations of "business" and "corporation" from critics of capitalism or corporate excess.

An example of an agribusiness was the Old North State Winegrowers Cooperative in North Carolina. Wine grape farmers came together to not only sell their grapes but to share a winery, winemaker and marketing brand together. The cooperative failed in 2006, three years after opening its winery.

Many progressive agribusinesses are now operating online businesses. Rising fuel costs are increasingly adding financial burdens on the day to day running of agricultural companies. An example of an online agribusiness is FarmingPages.com[1].

AGRICULTURAL SCIENCE

Agricultural science is a broad multidisciplinary field that encompasses the parts of exact, natural, economic and social sciences that are used in the practice and understanding of agriculture. (Veterinary science, but not animal science, is often excluded from the definition.)

Agriculture and Agricultural Science

The two terms are often confused. However, they cover different concepts:

- Agriculture is the set of activities that transform the environment for the production of animals and plants for human use. Agriculture concerns techniques, including the application of agronomic research.
- Agronomy is research and development related to studying and improving plant-based agriculture.

Agricultural sciences include research and development on:

- Production techniques (e.g., irrigation management, recommended nitrogen inputs)
- Improving agricultural productivity in terms of quantity and quality (e.g., selection of drought-resistant crops and animals, development of new pesticides, yield-sensing technologies, simulation models of crop growth, in-vitro cell culture techniques)
- Transformation of primary products into end-consumer products (e.g., production, preservation, and packaging of dairy products)
- Prevention and correction of adverse environmental effects (e.g., soil degradation, waste management, bioremediation)
- Theoretical production ecology, relating to crop production modeling
- traditional agricultural systems, sometime termed subsistence agriculture, which feed most of the poorest people in the world. These systems are of interest as they sometimes retain a level of integration with natural ecological greater than that of industrial agriculture, which in may be more sustainable than some modern agricultural systems
- Food production and demand on a global basis, with special attention paid to the major producers, such as China, India, Brazil and the USA.

Agricultural Science: a Local Science

With the exception of theoretical agronomy, research in agronomy, more than in any other field, is strongly related to local areas. It can be considered a science of ecoregions, because it is closely linked to soil properties and climate, which are never exactly the same from one place to another. Many people think an agricultural production system relying on local weather, soil characteristics, and specific crops has to be studied locally. Others feel a need to know and understand production systems in as many areas as possible, and the human dimension of interaction with nature.

History of Agricultural Science

Agricultural science began with Mendel's genetic work, but in modern terms might be better dated from the chemical fertilizer outputs of plant physiological understanding in eighteenth century Germany. In the United States, a scientific revolution in agriculture began with the Hatch Act of 1887, which used the term "agricultural science". The Hatch Act was driven by farmers' interest in knowing the constituents of early artificial fertilizer. The Smith-Hughes Act of 1917 shifted agricultural education back to its vocational roots, but the scientific foundation had been built. After 1906, public expenditures on agricultural research in the US exceeded private expenditures for the next 44 years.

Intensification of agriculture since the 1960s in developed and developing countries, often referred to as the Green Revolution, was closely tied to progress made in selecting and improving crops and animals for high productivity, as well as to developing additional inputs such as artificial fertilizers and phytosanitary products.

As the oldest and largest human intervention in nature, the environmental impact of agriculture in general and more recently intensive agriculture, industrial development, and population growth have raised many questions among agricultural scientists and have led to the development and emergence of new fields. These include technological fields that assume the solution to technological problems lies in better technology, such as integrated pest management, waste treatment technologies, landscape architecture, genomics, and agricultural philosophy fields that include references to food production as something essentially different from non-essential economic 'goods'. In fact, the interaction between these two approaches provide a fertile field for deeper understanding in agricultural science.

New technologies, such as biotechnology and computer science (for data processing and storage), and technological advances have made it possible to develop new research fields, including genetic engineering, agrophysics, improved statistical analysis, and precision farming.

Balancing these, as above, are the natural and human sciences of agricultural science that seek to understand the human-nature interactions of traditional agriculture, including interaction of religion and agriculture, and the non-material components of agricultural production systems.

Prominent Agricultural Scientists

- Justus von Liebig
- Robert Bakewell
- Norman Borlaug
- Luther Burbank
- George Washington Carver
- René Dumont
- Jay Lush
- Gregor Mendel
- Louis Pasteur
- Eli Whitney
- Sewall Wright

Agricultural Science and Agriculture Crisis

Agriculture sciences seek to feed the world's population while preventing biosafety problems that may affect human health and the environment. This requires promoting good management of natural resources and respect for the environment, and increasingly concern for the psychological wellbeing of all concerned in the food production and consumption system.

Economic, environmental, and social aspects of agriculture sciences are subjects of ongoing debate. Recent crises (such as Avian Flu, mad cow disease and issues such as the use of genetically modified organisms) illustrate the complexity and importance of this debate.

AGRONOMY

Agronomy is the science and technology of using plants for food, fuel, feed, and fiber. Agronomy encompasses work in the areas of plant genetics, plant physiology, meteorology, and soil science. Agronomy is the application of a combination of sciences like biology, chemistry, ecology, earth science, and genetics. Agronomists today are involved with many issues including producing food, creating healthier food, managing environmental impact of agriculture, and creating energy from plants. Agronomists often specialize in areas such as crop rotation, irrigation and drainage, plant breeding, soil classification, soil fertility, weed control, insect and pest control.

Plant Breeding

This area of agronomy involves selective breeding of plants to produce the best crops under various conditions. Plant breeding has increased crop yields and has improved the nutritional value of numerous crops, including corn, soybeans, and wheat. It also has led to the development of new types of plants. For example, a hybrid grain called triticale was produced by crossbreeding rye and wheat. Triticale contains more usable protein than does either rye or wheat. Agronomy has also been instrumental in fruit and vegetable production research. It is understood that the role of agronomist includes seeing whether produce from a field of 'x' meets the following conditions: 1. Land and water access, 2. Commercialisation (market), 3. Quality and quantity of inputs, 4. Risk protection (insurance), 5. Agricultural credit.

Biotechnology

Agronomists use biotechnology to extend and expedite the development of desired characteristics listed in the Plant Breeding section. Biotechnology is often a lab activity requiring field testing of the new crop varieties that are developed.

In addition to increasing crop yields, reducing crop vulnerability to environmental stresses, improving health and taste of foods, and reducing the need for field applied chemicals, agronomic biotechnology is increasingly being applied for novel uses other than food. For example, oilseed is at present used mainly for margarine and other food oils, but it can be modified to produce fatty acids for detergents, substitute fuels and petrochemicals.

Soil Science

Agronomists study sustainable ways to make soils more productive. They classify soils and reproduce them to determine whether they contain substances vital to plant growth such as compounds of nitrogen, phosphorus, and potassium. If a certain soil is deficient in these substances, fertilizers may provide them.

Soil science also involves investigation of the movement of nutrients through the soil, the amount of nutrients absorbed by a plant's roots, and the development of roots and their relation to the soil.

Soil Conservation

In addition, agronomists develop methods to preserve the soil and to decrease the effects of erosion by wind and water. For example, a technique called contour plowing may be used to prevent soil erosion and conserve rainfall.

Researchers in agronomy also seek ways to use the soil more effectively in solving other problems. Such problems include the disposal of human and animal wastes; water pollution; and the build-up in the soil of pesticides. No-tilling crops is a technique now used to help prevent erosion. planting of soil binding grasses along contours can be tried in steep slopes. For better effect, contour drains of depths up to 1 metre may help retain the soil and prevent permanent wash off.

Agroecology

Agroecology is the management of agricultural systems with an emphasis on ecological and environmental perspectives. This area is closely associated with work in the areas of sustainable agriculture, organic farming, alternative food systems and the development of alternative cropping systems.

THEORETICAL MODELLING

Agronomy Schools

Agronomy programs are offered at colleges, universities, and specialized agricultural schools. Agronomy programs often involve classes across a range of departments including agriculture, biology, chemistry, and physiology.

Career Outlook

Past agricultural research has created higher yielding crops, crops with better resistance to pests and plant pathogens, and more effective fertilizers and pesticides. Research is still necessary, however, particularly as insects and diseases continue to adapt to pesticides and as soil fertility and water quality continue to need improvement. Emerging biotechnologies will play an ever larger role in agricultural research. Scientists will be needed to apply these technologies to the creation of new food products and other advances. Moreover, increasing demand is expected for biofuels and other agricultural products used in industrial processes. Agricultural scientists will be needed to find ways to increase the output of crops used in these products.

Agronomists will also be needed to balance increased agricultural output with protection and preservation of soil, water, and ecosystems. They increasingly encourage the practice of sustainable agriculture by developing and implementing plans to manage pests, crops, soil fertility and erosion, and animal waste in ways that reduce the use of harmful chemicals and do little damage to farms and the natural environment.

Most agronomists are consultants, researchers, or teachers. Many work for agricultural experiment stations, federal or state government agencies, industrial firms, or universities. Agronomists also serve in such international organizations as the Agency for International Development and the Food and Agriculture Organization of the United Nations.

ANIMAL HUSBANDRY

Animal husbandry, also called animal science, stockbreeding or simple husbandry, is the agricultural practice of breeding and raising livestock. It has been practiced for thousands of years, since the first domestication of animals.

The science of animal husbandry is taught in many universities and colleges around the world. Students of animal science may pursue degrees in veterinary medicine following graduation, or go on to pursue master's degrees or doctorates in disciplines such as nutrition, genetics and animal breeding, or reproductive physiology. Graduates of these programs may be found working in the veterinary and human pharmaceutical industries, the livestock and pet supply and feed industries, farming or in academia.

Historically, certain sub-professions within the field of animal husbandry are specifically named according to the animals that are cared for.

Different Types of Animal Husbandry

A swineherd is a person who cares for hogs and pigs (older English term: swine). A shepherd is a person who cares for sheep. A goatherd cares for goats. A cowherd cares for cattle. In previous years, it was common to have herds which were made up of sheep and goats. In his case, the person tending them was called a shepherd. Camels are also cared for in herds. In Tibet yaks are herded. In Latin America, llamas and alpacas are herded.

In more modern times, the cowboys or *vaqueros* of North and South America ride horses and participate in cattle drives to watch over cows and bulls raised primarily for food. In Australia many herds are managed by farmers on motorbikes and in helicopters. Today, herd managers often oversee thousands of animals and many staff. Farms and ranches may employ breeders, herd health specialists, feeders, and milkers to help care for the animals. Techniques such as artificial insemination and embryo transfer are frequently used, not only as methods to guarantee that females are bred, but to help improve herd genetics. This may be done by transplanting embryos from stud-quality females, into flock-quality surrogate mothers-freeing up the stud-quality mother to be reimpregnated. This practice vastly increases the number of offspring which may be produced by a small selection of stud-quality parent animals. This in turn improves the ability of the animals to convert feed to meat, milk, or fiber more efficiently and improve the quality of the final

EXTENSIVE FARMING

Extensive farming is the use of manure as the only fertiler, including human and animal. Compost is sometimes used as well.(as opposed to intensive farming) is an agricultural production system that uses little inputs on vast areas of land, such as the Great Plains.

Extensive farming most commonly refers to sheep and cattle farming in areas with low agricultural productivity, but can also refer to large-scale growing of wheat, barley and other grain crops in areas like the Murray-Darling Basin. Here, owing to the extreme age and poverty of the soils, yields per hectare are very low, but the flat terrain and very large farm sizes mean yields per unit of labour are high. Nomadic herding is an extreme example of extensive farming, where herders move their animals to use feed from occasional rainfalls.

Geography

Extensive farming is found in the mid-latitude sections of most continents, as well as in desert regions where water for cropping is not available. The nature of extensive farming means it requires less rainfall than intensive farming. The farm is usually large in comparison with the numbers working and money spent on it. In most parts of Western Australia, pastures are so poor that only one sheep to the square mile can be supported. Just as the demand has led to the basic division of cropping and pastoral activities, these areas can also be subdivided depending on the regions rainfall, vegetation type and agricultural activity within the area and the many other parentheses related to this data.

Advantages

Extensive farming has a number of advantages over intensive farming:

1. Less labour per unit areas is required to farm large areas, especially since expensive alterations to land (like terracing) are completely absent.
2. Mechanisation can be used more effectively over large, flat areas.
3. Greater efficiency of labour means generally lower costs.
4. Animal welfare is generally improved because animals are not kept in stifling conditions.
5. Lower requirements of inputs such as fertilisers.
6. If animals are grazed on pastures native to the locality, there is less likely to be problems with exotic species.

Disadvantages

Extensive farming can have the following problems :

1. Yields tend to be much lower than with intensive farming
2. Animals are not fed continuously and growth is restricted (e.g. in droughts).
3. Large land requirements limit the habitat of wild species (in some cases, even very low stocking rates can be dangerous).
4. Susceptibity to the vagaries of climate can cause long-term ecological damage if drought or flood alters topography or erodes soils.

FACTORY FARMING

Factory farming is the practice of raising farm animals in confinement at high stocking density, where a farm operates as a factory — a practice typical in industrial farming by agribusinesses.

Confinement at high stocking density is one part of a systematic effort to produce the highest output at the lowest cost by relying on economies of scale, modern machinery, biotechnology, and global trade. Confinement at high stocking density requires antibiotics and pesticides to mitigate the spread of disease and pestilence exacerbated by these crowded living conditions. There are differences in the way factory farming techniques are practiced around the world. There is a continuing debate over the benefits and risks of factory farming. The issues include the efficiency of food production; animal welfare; whether it is essential for feeding the growing global human population; the environmental impact and the health risks.

The UN and OIE estimate that in coming decades there will be billions of additional consumers in developing countries eating meat factory farmed in developing countries, but currently only about 40 out of the around 200 countries in the world have the capacity to adequately respond to a health crisis originating from animal disease (such as avian flu, West Nile virus, bluetongue, and foot and mouth disease). Widespread use of antibiotics increases the chance of a pandemic resistant to known measures, which is exacerbated by a globally distributed food system. Decreased genetic diversity increases the chance of a food crisis.

TERMINOLOGY

Factory Farming

The Oxford English Dictionary attributes the first recorded use of "factory farming" to an American journal of economics in 1890. It is now used widely by mainstream news organizations, including the BBC, *The Washington Post*, and CNN. A 1998 documentary, *A Cow at My Table*, shows the term is also used within the agricultural industry, although it is regarded by sections of the industry as a term used by activists. The *Encyclopaedia Britannica* writes that the term is "descriptive of standard farming practice in the U.S." and frequently used by animal rights activists. *Webster's New Millennium* defines it as "a system of large-scale industrialized and intensive agriculture that is focused on profit with animals kept indoors and restricted in mobility."

Animal Feeding Operation (AFO)

"An animal feeding operation is defined by the United States Environmental Protection Agency as a lot or facility where animals are kept 45 days of the year or more [and] structures or animal traffic prevents vegetative growth. Note that this is different from a EPA's definition of a confined animal feeding operation which is an animal feeding operation larger than a given size."

Confined Animal Feeding Operations

In the U.S., some factory farms are also known as Confined Animal Feeding Operations (CAFOs), concentrated animal feeding operations, or intensive livestock operations (ILOs). "A *confined animal feeding operation* means a lot or facility, together with any associated treatment works, where both of the following conditions are met. One, animals have been, are, or will be stabled or confined and fed or maintained for a total of 45 days or more in any 12-month period. And two, crops, vegetation, forage growth, or post-harvest residues are not sustained over any portion of the operation lot or facility." The definition is used as part of waste management and environmental protection laws to deal with the concentrated pollution from large quantities of animal waste..

Confinement

CAFOs and factory farms can be mostly indoors or mostly outdoors operations. The "confinement at high stocking density" aspect refers to lack of natural vegetation that the animals can eat and that can naturally process the resulting animal waste. High stocking density destroys the vegetation and produces unacceptable pollution from the animal waste in run-off and ground water unless it is handled appropriately, so laws have been enacted to deal with that; thus the *legal* definition for the term CAFO. Caged for life in pens too small to be humane is a completely separate issue from what "confined" refers to when used to define "factory farms" and "CAFO"s.

History

Agriculture adopted more intensive methods during the 18th century, with this growth in production best characterised by the Agricultural Revolution, where improvements in farming techniques allowed for significantly improved yields, and supported the urbanisation of the population during the Industrial Revolution.

Innovations in agriculture beginning in the late 19th century paralleled developments in mass production in other industries. The identification of nitrogen and phosphorus as critical factors in plant growth led to the manufacture of synthetic fertilizers, making possible more intensive types of agriculture. The discovery of vitamins and their role in animal nutrition, in the first two decades of the 20th century, led to vitamin supplements, which in the 1920s allowed certain livestock to be raised indoors. The discovery of antibiotics and vaccines facilitated raising livestock in larger numbers by reducing disease. Chemicals developed for use in World War II gave rise to synthetic pesticides. Developments in shipping networks and technology have made long-distance distribution of agricultural produce feasible.

According to the BBC, factory farming in Britain began in 1947 when a new Agriculture Act granted subsidies to farmers to encourage greater output by introducing new technology, in order to reduce Britain's reliance on imported meat. The United Nations writes that intensification of animal production was seen as a way of providing food security. The agriculture correspondent of *The Guardian* wrote in 1964:

Factory farming, whether we like it or not, has come to stay. The tide will not be held back, either by the humanitarian outcry of well meaning but sometimes misguided animal lovers, by the threat implicit to traditional farming methods, or by the sentimental approach to a rural way of life. In a year which has been as uneventful on the husbandry side as it has been significant in economic and political developments touching the future of food procurement, the more far-seeing would name the growth of intensive farming as the major development.

NATURE OF THE PRACTICE

Scale

Agricultural production across the world doubled four times between 1820 and 1975 to feed a global population of one billion human beings in 1800 and 6.5 billion in 2002. During the same period, the number of people involved in farming dropped as the process became more automated. In the 1930s, 24 percent of the American population worked in agriculture compared to 1.5 percent in 2002; in 1940, each farm worker supplied 11 consumers, whereas in 2002, each worker supplied 90 consumers.

The number of farms has also decreased, and their ownership is more concentrated. In the U.S., four companies produce 81 percent of cows, 73 percent of sheep, 57 percent of pigs and 50 percent of chickens. In 1967, there were one million pig farms in America; as of 2002, there were 114,000, with 80 million pigs (out of 95 million) killed each year on factory farms as of 2002, according to the U.S. National Pork Producers Council. According to the Worldwatch Institute, 74 percent of the world's poultry, 43 percent of beef, and 68 percent of eggs are produced this way.

Although Europe has become increasingly skeptical of factory farming, after a series of diseases such as BSE (mad cow) and foot and mouth disease affected its agricultural industries, globally there are indications that the industrialized production of farm animals is set to increase. According to Denis Avery of the Hudson Institute, Asia increased its consumption of pork by 18 million tons in the 1990s. As of 1997, the world had a stock of 900 million pigs, which Avery predicts will rise to 2.5 billion pigs by 2050. He told the College of Natural Resources at the University of California, Berkeley that three billion pigs will thereafter be needed annually to meet demand.

Distinctive Characteristics

Factory farms hold large numbers of animals, typically cows, pigs, turkeys, or chickens, often indoors, typically at high densities. The aim of the operation is to produce as much meat, eggs, or milk at the lowest possible cost. Food is supplied in place, and a wide variety of artificial methods are employed to maintain animal health and improve production, such as the use of antimicrobial agents, vitamin supplements, and growth hormones. Physical restraints are used to control movement or actions regarded as undesirable. Breeding programs are used to produce animals more suited to the confined conditions and able to provide a consistent food product. The distinctive characteristic of factory farms is the intense concentration of livestock. At one farm (Farm 2105) run by Carrolls Foods of North Carolina, the second-largest pig producer in the U.S., twenty pigs are kept per pen and each confinement building or "hog parlor" holds 25 pens. The company's chief executive officer, F.J. "Sonny" Faison, has said: "It's all a supply-and-demand price question ... The meat business in this country is just about perfect, uncontrolled supply-and-demand free enterprise. And it continues to get more and more sophisticated, based on science. Only the least-cost producer survives in agriculture." The animals are better off in total confinement, according to Faison:

They're in state-of-the-art confinement facilities. The conditions that we keep these animals in are much more humane than when they were out in the field. Today they're in housing that is environmentally controlled in many respects. And the feed is right there for them all the time, and water, fresh water. They're looked after in some of the best conditions, because the healthier and [more] content that animal, the better it grows. So we're very interested in their well-being—up to an extent.

KEY ISSUES

Ethics

The large concentration of animals, animal waste, and the potential for dead animals in a small space poses ethical issues. It is recognised that some techniques used to sustain intensive agriculture can be cruel to animals. As awareness of the problems of intensive techniques has grown, there have been some efforts by governments and industry to remove inappropriate techniques. In the UK, the Farm Animal Welfare Council was set up by the government to act as an independent advisor on animal welfare in 1979. and expresses its policy as five freedoms: from hunger & thirst; from discomfort; from pain, injury or disease; to express normal behaviour; from fear and distress.

There are differences around the world as to which practices are accepted and there continue to be changes in regulations with animal welfare being a strong driver for increased regulation. For example, the EU is bringing in further regulation to set maximum stocking densities for meat chickens by 2010, where the UK Animal Welfare Minister commented, "The welfare of meat chickens is a major concern to people throughout the European Union. This agreement sends a strong message to the rest of the world that we care about animal welfare."

However, given the assumption that intensive farming techniques are a necessity, it is recognized that some apparently cruel techniques are better than the alternative. For example, in the UK, de-beaking of chickens is deprecated, but it is recognized that it is a method of last resort, seen as better than allowing vicious fighting and ultimately cannibalism. Between 60 and 70 percent of six million breeding sows in the U.S. are confined during pregnancy, and for most of their adult lives, in 2 ft (0.61 m) by 7 ft (2.1 m) gestation crates. According to pork producers and many veterinarians, sows will fight if housed in pens. The largest pork producer in the U.S. said in January 2007 that it will phase out gestation crates by 2017. They are being phased out in the European Union, with a ban effective in 2013 after the fourth week of pregnancy. With the evolution of factory farming, there has been a growing awareness of the issues amongst the wider public, not least due to the efforts of animal rights and welfare campaigners. As a result gestation crates, one of the more contentious practices, are the subject of laws in the U.S., Europe and around the world to phase out their use as a result of pressure to adopt less confined practices.

U.S. Consumer Preferences

A study done by Oklahoma State University, utilizes results of a representative telephone survey to measure consumer attitudes towards farm animal welfare, and investigates how these attitudes vary across individuals. The survey finds that consumers desire high standards of animal care, even if it raises food prices and involves government regulation. Support is particularly strong for females, Democrats, and residents of the Northeastern United States. To provide high standards of animal care, consumers as a whole perceive allowing animals to exhibit natural behaviors and exercise outdoors to be more important than protection from other animals, shelter, socialization, and comfortable bedding. Consumers vary in their perceptions though, and are divided into three classes of consumers: Naturalists, Price Seekers, and Descartes' Entourage. Naturalists place great importance on allowing animals to exhibit natural behaviors and exercise outdoors, and comprise 46% of the sample. Price Seekers, comprising 14% of the sample, are primarily concerned with low prices. Descartes' Entourage make up 40% of the respondents, and value animal welfare but perceive it can be achieved by simply providing food, water, and treatment for injury and disease. This last group perceives amenities such as access to outdoors and ability to exhibit natural behaviors unimportant for the well-being of farm animals.

Health Problems and Nuisance

According to the U.S. Centers for Disease Control and Prevention (CDC), farms on which animals are intensively reared can cause adverse health reactions in farm workers. Workers may develop acute and chronic lung disease, musculoskeletal injuries, and may catch infections that transmit from animals to human beings.

Pesticides are used to control organisms which are considered harmful and they save farmers money by preventing product losses to pests. In the US, about a quarter of pesticides used are used in houses, yards, parks, golf courses, and swimming pools and about 70% are used in agriculture. However, pesticides can make their way into consumers' bodies which can cause health problems. One source of this is bioaccumulation in animals raised on factory farms.

The CDC writes that chemical, bacterial, and viral compounds from animal waste may travel in the soil and water. Residents near such farms report nuisances such as unpleasant smell, flies and adverse health effects.

The CDC has identified a number of pollutants associated with the discharge of animal waste into rivers and lakes, and into the air. The use of antibiotics may create antibiotic-resistant pathogens; parasites, bacteria, and viruses may be spread; ammonia, nitrogen, and phosphorus can reduce oxygen in surface waters and contaminate drinking water; pesticides and hormones may cause hormone-related changes in fish; animal feed and feathers may stunt the growth of desirable plants in surface waters and provide nutrients to disease-causing micro-organisms; trace elements such as arsenic and copper, which are harmful to human health, may contaminate surface waters.

In the European Union, growth hormones are banned on the basis that there is no way of determining a safe level. The UK has stated that in the event of the EU raising the ban at some future date, to comply with a precautionary approach, it would only consider the introduction of specific hormones, proven on a case by case basis. The various techniques of factory farming have been associated with a number of European incidents where public health has been threatened or large numbers of animals have had to be slaughtered to deal with disease. Where disease breaks out, it may spread more quickly, not only due to the concentrations of animals, but because modern approaches tend to distribute animals more widely.. The international trade in animal products increases the risk of global transmission of virulent diseases such as swine fever, BSE, foot and mouth and bird flu.

Aspects of Factory Farming

- Low monetary cost — Intensive agriculture tends to produce food that can be sold at lower cost to consumers. This is achieved by reducing land costs and management costs.
- Standardization — Factory farming methods permit increased consistency and control over product output.
- Efficiency — Animals in confinement can be supervised more closely than free-ranging animals, and diseased animals can be treated faster. Further, more efficient production of meat, milk, or eggs results in a need for fewer animals to be raised-at the expense of food quality and animal or plant health.
- Economic contribution — The high input costs of agricultural operations result in a large influx and distribution of capital to a rural area from distant buyers

rather than simply recirculating existing capital. A single dairy cow contributes over $1300 US to a local rural economy each year, each beef cow over $800, meat turkey $14, and so on. As Pennsylvania Secretary of Agriculture Dennis Wolff states, "Research estimates that the annual economic impact per cow is $13,737. In addition, each $1 million increase in PA milk sales creates 23 new jobs. This tells us that dairy farms are good for Pennsylvania's economy."

- Food safety — Reducing number and diversity of agricultural production facilities may make oversight and regulation of food quality easier. Centralized and standardized food production can protect general food safety.
- Animal health — Larger farms may employ greater resources to maintain a high level of animal health. Larger farms can potentially employ expert employees who devote their working hours to assessing animal health, a task which would be cost-prohibitive for most small farms. Larger farms may be more able to make regular use of veterinarians and the resources of state and federal agricultural extension services. Industrial agriculture generally provides more mechanisms for the use of antibiotics to prevent and treat diseases than non-industrial agriculture.
- Diseases-Intensive farming may make the evolution and spread of harmful diseases easier. Many communicable diseases spread rapidly through densely spaced populations of animals with low genetic diversity. Animals raised on antibiotics may develop antibiotic resistant strains of pathogenic bacteria ("superbugs"). Use of animal vaccines can create new viruses that kill people and cause flu pandemic threats. H5N1 is an example of where this might have already occurred.
- Pollution — Large quantities and concentrations of waste are produced. Lakes, rivers, and groundwater are at risk when animal waste is improperly recycled. Pollutant gases are also emitted. Concentrations of animals can

produce unacceptable levels of foul smells as opposed to the tolerable odours of the countryside. In less intensive conditions, natural processes can break down potential pollutants. Large farms can maintain and operate sophisticated systems to control waste products. Smaller farms may be less able to invest in the same standards of pollution control.

- Ethics — Cruelty to animals: Crowding, drugging, and performing surgery on animals. On some farms, chicks may be debeaked when very young. Confining hens and pigs may lead to physical problems such as osteoporosis and joint pain, and psychological problems including boredom and frustration, as shown by repetitive or self-destructive actions. Animal treatment is subject to welfare legislation, though there is not consensus on what is acceptable. Some harmful treatments, such as debeaking, are tolerated on the basis that the alternative is greater harm to the animals.
- Destruction of biodiversity — A tendency towards using single adapted breeds (a mono-culture) in factory farming, both in arable and animal farming, gives uniform product designed for high yields, at the risk of increased susceptibility to disease. The loss of locally adapted breeds reduces the resilience of the agricultural system. The issue is not limited to factory farming and historically the problem is reflected in the rapid adoption of one or two strains of crops across a wide area as seen in the Irish potato famine of 1845 and the Bengal rice famine in 1942. The loss of the gene pool of domesticated animals limits the ability to adapt to future problems. This issue exists in all types of farming practices.
- Nutrient poor food-Heavy reliance on non-heritage, high yield breeds of plants and animals reduce nutrient content. This is also symptomatic of heavy use of heavy chemical pesiticides, fertilizers and depleted soil conditions from intensive, non-rotated use.

FREE RANGE

Free range is a method of farming husbandry where the animals are allowed to roam freely instead of being contained in any manner. The term is used in two senses that do not overlap completely: as a farmer-centric description of husbandry methods, and as a consumer-centric description of them. Farmers practice free range to achieve free-range or humane certification (and thus capture high prices), to reduce feed costs, to improve the happiness and liveliness of their animals, to produce a higher-quality product, as a method of raising multiple crops on the same land, or for other reasons.

In practice, there are few regulations imposed on what can be called "free range," and the term may be used misleadingly to imply that the animal product has been produced more humanely than it actually has been.

Free range may apply to meat, eggs or dairy farming.

In ranching, free-range livestock are permitted to roam without being fenced in, as opposed to fenced-in pastures. In many of the agriculture based economies, free-range livestock are quite common.

Salmonella infection rates in free-range and organic chickens have been found to be comparable to those produced in typical poultry production houses.

History

If one allows "free range" to include "herding", free range was a typical husbandry method at least until the development of barbed wire and chicken wire. The generally poor understanding of nutrition and diseases before the Twentieth Century made it difficult to raise many livestock species without giving them access to a varied diet, and the labor of keeping livestock in confinement and carrying all their feed to them was prohibitive except for high-profit animals such as dairy cattle.

In the case of poultry, free range was the dominant system until the discovery of vitamins A and D in the 1920s allowed confinement to be practiced successfully on a commercial scale.

Before that, green feed and sunshine (for the vitamin D) were necessary to provide the necessary vitamin content. Some large commercial breeding flocks were reared on pasture into the 1950s. Nutritional science resulted in the increased use of confinement for other livestock species in much the same way.

United States

Traditional American usage equates "free-range" with "unfenced," and with the implication that there was no herdsman keeping them together or managing them in any way. Legally, a free-range jurisdiction allowed livestock (perhaps only of a few named species) to run free, and the owner was not liable for any damage they caused. In such jurisdictions, people who wished to avoid damage by livestock had to fence them out; in others, the owners had to fence them in.

In recent years, with the days of free-range cattle mostly in the past, neither the presence of a "legal fence" surrounding the farm nor the pros and cons of old-time free-range ranching are the main points of interest. Instead, the term "free range" is mainly used as a marketing term rather than a husbandry term, meaning something on the order of, "low stocking density," "pasture-raised," "grass-fed," "old-fashioned," "humanely raised," etc. In poultrykeeping, "Free range" is widely confused with yarding, which means keeping poultry in fenced yards. In reality, the two methods have little in common.

The U.S. Department of Agriculture requires that chickens raised for their meat have access to the outdoors in order to receive the free-range certification. Free-range chicken eggs, however, have no legal definition in the United States. Likewise, free-range egg producers have no common standard on what the term means. Many egg farmers sell their eggs as free range merely because their cages are two or three inches above average size, or because there is a window in the shed.

The USDA has no specific definition for "free-range" beef, pork, and other non-poultry products. All USDA definitions of "free-range" refer specifically to poultry. No other criteria-such as the size of the range or the amount of space given to each animal-are required before beef, lamb, and pork can be called "free-range". Claims and labeling using "free range" are therefore unregulated. The USDA relies "upon producer testimonials to support the accuracy of these claims."

In a December 30, 2002 Federal Register notice and request for comments (67 Fed. Reg. 79552), USDA's Agricultural Marketing Service proposed "minimum requirements for livestock and meat industry production/marketing claims". Many industry claim categories are included in the notice, including breed claims, antibiotic claims, and grain fed claims. "Free Range, Free Roaming, or Pasture Raised" would be defined as "livestock that have had continuous and unconfined access to pasture throughout their life cycle" with an exception for swine ("continuous access to pasture for at least 80% of their production cycle"). This proposed rulemaking is still in play.

In a May 12, 2006 Federal Register notice (71 Fed. Reg. 27662), the agency presented a summary and its responses to comments received in the 2002 notice, but only for the category "grass (forage) fed" which the agency stated was to be a category separate from "free range". Comments received for other categories, including "free range," are to be published in future Federal Register editions.

The broadness of "free range" in the U.S. has caused some people to look for alternative terms. "Pastured poultry" is a term promoted by farmer/author Joel Salatin for broiler chickens raised on grass pasture for all of their lives except for the initial brooding period. The Pastured Poultry concept is promoted by the American Pastured Poultry Producers' Association (APPPA), an organization of farmers raising their poultry using Salatin's principles.

Alternative terminology can also be used to make high-density confinement sound more palatable. For example: *cage-free, free-running, free-roaming, naturally nested,* etc. are used as an alternative to the technical term, *high-density floor confinement.* Whether high-density floor confinement is more humane than high-density cage confinement is arguable, but in any event high-density confinement (of whatever type) is the antithesis of free range.

European Union

The European Union regulates marketing standards for egg farming which specifies the following (cumulative) minimum conditions for Free range method:

- hens have continuous daytime access to open-air runs, except in the case of temporary restrictions imposed by veterinary authorities,
- the open-air runs to which hens have access is mainly covered with vegetation and not used for other purposes except for orchards, woodland and livestock grazing if the latter is authorised by the competent authorities,
- the open-air runs must at least satisfy the conditions specified in Article 4(1)(3)(b)(ii) of Directive 1999/74/EC whereby the maximum stocking density is not greater than 2 500 hens per hectare of ground available to the hens or one hen per $4m^2$ at all times and the runs are not extending beyond a radius of 150 m from the nearest pophole of the building; an extension of up to 350 m from the nearest pophole of the building is permissible provided that a sufficient number of shelters and drinking troughs within the meaning of that provision are evenly distributed throughout the whole open-air run with at least four shelters per hectare.

Otherwise, egg farming in EU is classified into 4 categories: Organic (ecological), Free range, Barn, and Cages, each category being progressively worse for the well-being of the animal.) The mandatory labelling on the egg shells attributes a number (which is the first digit on the label) to each of these categories: 0 for Organic, 1 for Free range, 2 for Barn and 3 for Cages.

United Kingdom

In the UK, the Department for Environment, Food and Rural Affairs says that a free-range chicken must have daytime access to open-air runs during at least half of their life. Unlike in the United States, this definition also applies to egg-laying hens.

INTENSIVE FARMING

Intensive farming or intensive agriculture is an agricultural production system characterized by the high inputs of capital, fertilizers, labour, or labour-saving technologies such as pesticides relative to land area.

This is in contrast to the concept of Extensive Agriculture which involves a low input of materials and labour with the crop yield depending largely on the naturally available soil fertility, water supply or other land qualities. Modern day forms of intensive crop based agriculture involve the use of mechanical ploughing, chemical fertilizers, herbicides, fungicides, insecticides, plant growth regulators and/or pesticides. It is associated with the increasing use of agricultural mechanization, which have enabled a substantial increase in production.

Intensive animal farming practices can involve very large numbers of animals raised on limited land which require large amounts of food, water and medical inputs (required to keep the animals healthy in cramped conditions). Very large or confined indoor intensive livestock operations (particularly descriptive of common US farming practices) are often referred to as Factory farming and are criticised by opponents for the low level of animal welfare standards and associated pollution and health issues.

Advantages

Intensive agriculture has a number of benefits:

- Significantly increased yield per acre relative to extensive farming.
- Often leads to cheaper priced products because of better general production rate for the cost of raw materials.
- Not much space for the animal(s) to move therefore less energy used up; so less food supplied to the cattle, which leads to cheaper products.
- The animals are being fed continuously for maximum growth, and they have very little exercising space to burn off the extra meat.

Disadvantages

Intensive farming alters the environment in many ways.

- Limits the natural habitat of some wild creatures and can lead to soil erosion.
- Use of fertilizers can alter the biology of rivers and lakes. Some environmentalists attribute the hypoxic zone in the Gulf of Mexico as being encouraged by nitrogen fertilization of the algae bloom.
- Pesticides can kill useful insects as well as those that destroy crops.
- Generally not sustainable.
- Often results in an inferior product.
- Use of chemicals on fields creates run-off, excess runs off into rivers and lakes causing pollution.

Pre Modern Intensive Farming

Pre modern intensive farming techniques and structures include terracing, rice paddies, and various forms of aquaculture.

Oysters

"Oysters were likely the first sea animal to be transported from one area to another and cultivated as food. The ancient world, while knowing little about the reproduction of oysters, knew much about the conditions necessary for their growth.

Pliny the Elder, a noted Roman naturalist of the first century, has left an account of artificial oyster beds established in Lake Lucrinus near Naples by a Sergius Orata about 95 B.C. Orata's methods consisted of preparing the grounds by removing other forms of marine life, planting seed oysters, cultivating the oysters by keeping them separated in order to grow to a well-formed, mature size, and finally harvesting them when they were ready for market.

Modern oyster farming, based on the knowledge of oyster biology, basically follows the Roman procedure.Fisheries and Oceans Canada] article *American Oyster*

Terrace

In agriculture, a terrace is a leveled section of a hilly cultivated area, designed as a method of soil conservation to slow or prevent the rapid surface runoff of irrigation water. Often such land is formed into multiple terraces, giving a stepped appearance. The human landscapes of rice cultivation in terraces that follow the natural contours of the escarpments like contour ploughing is a classic feature of the island of Bali and the Banaue Rice Terraces in Benguet, Philippines. In Peru, the Inca made use of otherwise unusable slopes by drystone walling to create terraces.

Rice Paddy

A paddy field is a flooded parcel of arable land used for growing rice and other semiaquatic crops. Paddy fields are a typical feature of rice-growing countries of east and southeast Asia including Malaysia, China, Sri Lanka, Myanmar, Thailand, Korea, Japan, Vietnam, Taiwan, Indonesia, India, and the Philippines. They are also found in other rice-growing regions such as Piedmont (Italy), the Camargue (France) and the Artibonite Valley (Haiti). They can occur naturally along rivers or marshes, or can be constructed, even on hillsides, often with much labour and materials. They require large quantities of water for irrigation, which can be quite complex for a highly developed system of paddy fields. Flooding provides water essential to the growth of the crop. It also gives an environment favourable to the strain of rice being grown, and is hostile to many species of weeds. As the only draft animal species which is adapted for life in wetlands, the water buffalo is in widespread use in Asian rice paddies. There are significant adverse environmental impacts from rice paddy cultivation due to the generation of large quantities of methane gas. World methane production due to rice paddies has been estimated in the range of 50 to 100 million tonnes per annum; this level of greenhouse gas generation is a large component of the global warming threat and derives simply from an expanding human population.

Rice-farming and the use of paddies in Korea is ancient. Korean paddy-farming can provide cultural background on the use of paddies in East Asia. A pit-house at the Daecheon-ni site yielded carbonized rice grains and radiocarbon dates indicating that rice cultivation may have begun as early as the Middle Jeulmun Pottery Period (c. 3500-2000 B.C.) in the Korean Peninsula (Crawford and Lee 2003). The earliest rice cultivation in the Korean Peninsula may have used dry-fields instead of paddies. The earliest Mumun features were usually located in low-lying narrow gulleys that were naturally swampy and fed by the local stream system. Some Mumun paddies in flat areas were made of a series of squares and rectangles separated by bunds approximately 10 cm in height, while terraced paddies consisted of long irregularly shapes that followed natural contours of the land at various levels (Bale 2001; Kwak 2001).

Mumun Period rice farmers used all of the elements that are present in today's paddies such terracing, bunds, canals, and small reservoirs. We can grasp some paddy-farming techniques of the Middle Mumun (c. 850-550 B.C.) from the well-preserved wooden tools excavated from archaeological rice paddies at the Majeon-ni Site. However, iron tools for paddy-farming were not introduced until sometime after 200 B.C. The spatial scale of individual paddies, and thus entire paddy-fields, increased with the regular use of iron tools in the Three Kingdoms of Korea Period (c. A.D. 300/400-668).

Modern Intensive Farming Types

Modern intensive farming refers to the industrialized production of animals (livestock, poultry and fish) and crops. The methods deployed are designed to produce the highest output at the lowest cost; usually using economies of scale, modern machinery, modern medicine, and global trade for financing, purchases and sales. The practice is widespread in developed nations, and most of the meat, dairy, eggs, and crops available in supermarkets are produced in this manner.

Sustainable Intensive Farming

Biointensive agriculture focuses on maximizing efficiency: yield per unit area, yield per energy input, yield per water input, etc. Agroforestry combines agriculture and orchard/

forestry technologies to create more integrated, diverse, productive, profitable, healthy and sustainable land-use systems. Intercropping can also increase total yields per unit of area or reduce inputs to achieve the same, and thus represents (potentially sustainable) agricultural intensification. Unfortunately, yields of any specific crop often diminish and the change can present new challenges to farmers relying on modern farming equipment which is best suited to monoculture. Vertical farming, a type of intensive crop production that would grow food on a large scale in urban centers, has been proposed as a way to reduce the negative environmental impact of traditional rural agriculture.

Intensive Aquaculture

Aquaculture is the cultivation of the natural produce of water (fish, shellfish, algae, seaweed and other aquatic organisms). Intensive Aquaculture can often involve tanks or other highly controlled systems which are designed to boost production for the available volume or area of water resource.

Intensive Livestock Farming

Main Article: Factory Farming

The modern examples of intensive farming are broadly referred to as Concentrated Animal Feeding Operations (CAFOs) or often termed Factory farming. These include:

- Intensive pig farming or Intensive piggery farming
- Large scale chicken farms
- Cattle feed lots

Managed Intensive Grazing

This sustainable intensive livestock management system is increasingly used to optimize production within a sustainability framework and is generally not considered Factory farming.

Individual Industrial Agriculture Farm

Major challenges and issues faced by individual industrial agriculture farms include:

- integrated farming systems
- crop sequencing
- water use efficiency
- nutrient audits
- herbicide resistance
- financial instruments (such as futures and options)
- collect and understand own farm information;
- knowing your products
- knowing your markets
- knowing your customers
- satisfying customer needs
- securing an acceptable profit margin
- cost of servicing debt
- ability to earn and access off-farm income;
- management of machinery and stewardship investments.

Integrated Farming Systems

An integrated farming system is a progressive biologically integrated sustainable agriculture system such as Integrated Multi-Trophic Aquaculture or Zero waste agriculture whose implementation requires exacting knowledge of the interactions of numerous species and whose benefits include sustainability and increased profitability.

Elements of this integration can include:

- intentionally introducing flowering plants into agricultural ecosystems to increase pollen-and nectar-resources required by natural enemies of insect pests
- using crop rotation and cover crops to suppress nematodes in potatoes

Crop Rotation

Crop rotation or crop sequencing is the practice of growing a series of dissimilar types of crops in the same space in sequential seasons for various benefits such as to avoid the build up of pathogens and pests that often occurs when one species is continuously cropped. Crop rotation also seeks to balance the fertility demands of various crops to avoid excessive depletion of soil nutrients. A traditional component of crop rotation is the replenishment of nitrogen through the use of green manure in sequence with cereals and other crops. It is one component of polyculture. Crop rotation can also improve soil structure and fertility by alternating deep-rooted and shallow-rooted plants.

Water use Efficiency

Crop irrigation accounts for 70% of the world's fresh water use. The agricultural sector of most countries is important both economically and politically, and water subsidies are common. Conservation advocates nave urged removal of all subsidies to force farmers to grow more water-efficient crops and adopt less wasteful irrigation techniques.

For crop irrigation and plant irrigation, optimal water efficiency means minimizing losses due to evaporation or runoff. An evaporation pan can be used to determine how much water is required to irrigate the land. Flood irrigation, the oldest and most common type, is often very uneven in distribution, as parts of a field may receive excess water in order to deliver sufficient quantities to other parts. Overhead irrigation, using center-pivot or lateral-moving sprinklers, gives a much more equal and controlled distribution pattern, but in extremely dry conditions much of the water may evaporate before it reaches the ground. Drip irrigation is the most expensive and least-used type, but offers the best results in delivering water to plant roots with minimal losses.

As changing irrigation systems can be a costly undertaking, conservation efforts often concentrate on maximizing the efficiency of the existing system. This may include chiseling compacted soils, creating furrow dikes to prevent runoff, and using soil moisture and rainfall sensors to optimize irrigation schedules.

Water catchment management measures include recharge pits, which capture rainwater and runoff and use it to recharge ground water supplies. This helps in the formation of ground water wells etc. and eventually reduces soil erosion caused due to running water.

Nutrient Audits

Better nutrient audits allow farmers to spend less money on nutrients and to create less pollution since less nutrient is added to the soil and thus there is less to run off and pollute. Methodologies for assessing soil nutrient balances have been studied and used for farms and entire countries for decades. But at present "there is no standard methodology for calculating nutrient budgets and there are no accepted 'benchmarks' figures against which to assess farm nutrient use efficiency. [A standard methodology] for calculating nutrient budgets on farms [is hoped to help reduce] diffuse water and air pollution from agriculture [through] best management practices in the use of fertilisers and organic manures, as part of the continued development of economically and environmentally sustainable farming systems."

Herbicide Resistance

In agriculture large scale and systematic weeding is usually required, often performed by machines such as cultivators or liquid herbicide sprayers. Selective herbicides kill specific targets while leaving the desired crop relatively unharmed. Some of these act by interfering with the growth of the weed and are often based on plant hormones. Weed control through herbicide is made more difficult when the weeds become resistant to the herbicide. Solutions include:

- using cover crops (especially those with allelopathic properties) that out-compete weeds and/or inhibit their regeneration.
- using a different herbicide
- using a different crop (e.g. genetically altered to be herbicide resistant; which ironically can create herbicide resistant weeds through horizontal gene transfer)
- using a different variety (e.g. locally-adapted variety that resists, tolerates, or even out-competes weeds)

- ploughing
- ground cover such as mulch or plastic
- manual removal

Agriculture plays a critical role in rural employment and environment management (Philip et al, 2005) and will face many challenges over the next century. Sustainable land development practices are urgently needed all over the world to preserve the production potential of agricultural land while protecting the environment (FAO, 1993). An integrated approach for sustainable agricultural development requires a combination of agricultural status, surface and groundwater potential, soils, natural vegetation, terrain characteristics, wetland etc. with meteorological information and socio-economic conditions. According to different experts, within the next decade, conventional field surveys will be phased out due to prohibiting costs and likely to be replaced by tools of Information Technology and remote sensing (Bourrough, 1986) to meet the increased demand for up-to-date and reliable information.

Agriculture has always been India's most important economic sector. Large area yield forecasting prior to harvest is of interest to government agencies, commodity firms and producers. Early information on yield and production helps in such activities as planning the type of crops (wheat, rice, barley, oats, etc.). Timely monitoring during the growing season provides opportunities to mitigate any revealed challenges. Information Technology, Remote Sensing Applications, and Decision Support Systems will facilitate decision makers' and farmers' knowledge of the state and situation of the agricultural system, thus helping them plan and act accordingly, Moni, 2003). Information and Communication Technology is a way to link the farmers to Universities and Research Centers, and researchers can orient their work to solve local problems in agriculture (Punjab Agricultural University, 1998).

Some forms of Agriculture Decision Support Systems have been discussed or implemented in different parts of the world. The Crop Condition Assessment Program (CCAP), (Stan Aronoff 2005) is a GIS based analytical Web application that shows in near real-time crop conditions within the Canadian prairies and the US. The MARS (Monitoring Agriculture with Remote Sensing) project (MARS Stat, 2004) focuses mainly on predicting the production volumes of the major crops at a national level and at a regional level for all EU member states. Their methodology is based on remote sensing, agro-meteorological data and crop growth model. The 'Global Monitoring for Environment and Security' (GMES) program is a European initiative and has a component which aims at supporting and complementing existing regional information and early warning systems on food and agriculture. The Global Information and Early Warning System on Food and Agriculture operated by the United Nations Food and Agriculture Organization (FAO) monitors famine-prone areas of the globe in order to predict food shortages and possible famine conditions. The Production Estimates and Crop Assessment Division (PECAD) of the US Department of Agriculture (USDA) relies on several different data sources to monitor weather anomalies that affect crop production and quality of agricultural commodities (Tetrault and Baldwin,2006). The main agro-meteorological input data sources are the ground meteorological station measurements, and grid weather data integrated with satellite imagery.

In addition, several crop models and data reduction algorithms are using both the station and satellite agro-meteorological data sets. These models include crop calendars, crop hazards, and several different crop yield reduction models to assess crop conditions. In India, the concept of an Agricultural Resource Information System (AgRIS) has been extensively discussed (Moni, 2003). The author expects the AgRIS Project to emerge as the richest "agricultural resources information system" covering all aspects of agriculture, natural resources, and food systems, linking the farmers to Universities and Research Centres, whereby researchers can orient their studies to solve local problems in agriculture.

AN AGRICULTURAL GEOCAPACITY CENTER FOR INDIA

In order to develop an Agricultural GeoCapacity Center for India, the AGCN project has been initiated by an Indo-

Canadian partnership. With the participation of the Canadian International Development Agency (CIDA), IES and PCI represent the Canadian team, while Punjab Remote Sensing Center (PRSC) and PAU (Punjab Agricultural University) represent the Indian counterpart.

Punjab, the home of the 'Green Revolution' in India, has been selected as a pilot area for the AGCN. After several discussion sessions organised in Ludhiana with PRSC and scientists at PAU, the following list of "burning issues" or main concerns for Punjab agriculture were identified:

1. Declining water table and extensive use of rice-wheat cropping system.
2. Declining soil health and increasing soil-water pollution.
3. Improvement of agro-meteorological weather forecasts. Maximization of rainfall agriculture and crop diversification.
4. Better use of rice-wheat straw and protection of environment.
5. Improvements in soil-water-nutrient management.
6. Integrated insect-pest-disease management.
7. Enhancement in crop productivity, quality of produce and sustainability of productivity.
8. Improvement of socio-economic status of Punjab farmers including issues of marketing, mechanization, land parcel size and others.
9. Land use changes and diversification of cropping patterns/farming systems.
10. Development of post-harvest technologies and value additions through processing.
11. Development/training of scientific human resources.

Based on these requirements, AGCN has been designed as an ensemble of tools and applications to provide information for:

- Optimisation of water use for irrigation:
- Precise weather forecasts

- Maximization of rainfall agriculture and crop diversification
- Better planning of:
- Land use
- Farm input, storage and output
- Better modeling and analysis of the crop system based on a multitude of inputs including weather, soil characteristics, population needs, and other economical market parameters
- Guidelines and strategic decisions based on local geographical, industrial and socio-economical characteristics of the area.The proposed Agricultural GeoCapacity Center Network project entails the establishment of a nation-wide information technology system for collecting, collating, organizing and interpreting information on land and resources collection and managing socio-economic data.

The proposed AGCN system, implements the above recommendations. It has enterprise architecture with core GeoCapacity services and data management. The enterprise architecture meets requirements, such as data distribution for efficient sharing and storage, services-oriented architecture for integrating best-of-breed technologies, multi-tier architecture for efficient and effective system enhancements, and environment componentization for system extensibility and scalability. The core of GeoCapacity services and data management provides fully automatic satellite data processing, centralized spatial data storage and management, and the interoperability required to access and use data and information from the various sources.

The AGCN will provide optimized performance with a common platform for advanced geospatial analysis. A job control system will manage the creation, execution, and monitoring of the fully automated processes. Alternative tools and algorithms for image classification will also be provided. Users will be able to access services through the Intranet or Internet, and can select various processing algorithms (such as image registration, image mosaicking, data loading, and map generation) and output formats (such as Web-based or printed maps). Also, various agriculture-specific applications can easily be built on top of the GeoCapacity core and services.

The AGCN is a modular system. Based on user gathered requirements, a number of modules have been identified within AGCN: Weather Application Module, Water Management Module, Cropping System Analysis Module, Soil Management Module, Pest and Disease Module, as well as Socio-Economic Module. The modules will be implemented as one or several applications for crop monitoring, crop yield forecasting, weather forecasting, etc. The applications are built around services provided mainly by the GIS core, like Image Processing Service, Map Generation Service, Web System Service, and Database Connection Service. The GIS core is the heart of the AGCN upon which everything is built. The GIS core provides services to build custom applications, and GIS desktop provides access to the database management system. The database management system will store, modify, and extract geospatial and non-geospatial information from the database.

AGCN APPLICATION MODULES

The next section briefly discusses the main components of the system: Weather applications, crop system and water management modules.

Weather Application Module

The weather application module consists of the following parts:

- Acquisition, checks and processing of daily meteorological station data.
- Spatial interpolation
- Production of weather indicators and daily Web-based maps
- Production of meteorological inputs for the statistical and dynamical crop models
- Processing of the land use map and management practices data and use of water budget and crop water

requirements models, production of seasonal map products for drought analysis and water deficit.

Daily meteorological station data and weather forecasts are used in different ways for crop yield evaluations. The information provided by this module is first used as weather indicators for direct evaluation of alarming situations such as droughts, extreme rainfall during different crop development phases of sowing, flowering and harvest; the second as input for the crop growth model and statistical yield model (MARS stat, 2004). The main objectives pursued with the weather module are:

- To evaluate the effects of weather on crops yields
- To produce inputs for the crop growth model for the monitoring and simulation of the crop behaviors.

Crop System Module

The crops' behaviors are mainly influenced by the atmospheric conditions near the earth's surface. The main purpose of the crop system module is to provide:

- Seasonal crop yield forecasts using statistical and dynamical crop models.
- Cropping patterns and crop rotation maps.
- Indices to characterize long term changes in the cropping system. Various indices, such as Multiple Cropping Index (MCI), Area Diversity Index (ADI), and Cultivated Land Utilization Index (CLUI) will be generated.

The cropping system analysis module produces estimated crop indicators like yield and production maps, cropping pattern maps and indices on performance of the cropping patterns. Different activities are expected in this module, some of the tasks like model validation will be done once at the start of the project and every five years:

1. Collection and processing of input data
2. Verification and validation of statistical model and dynamical models for yield forecast.
3. Methodology for crop acreage estimation

4. Spatial schematization (at grid level or agro-climatic zone)
5. Crop simulation
6. Spatial aggregation and production of crop indicators maps

The module uses daily interpolated grid weather to simulate biomass accumulation, crop development and yield forecast. Besides crop monitoring, the module will produce crop indicator maps and generate alarm warnings in case of abnormal conditions. The crop growth model is a complex process which takes place on farms at field level. Crop yields vary among regions, farms, fields, and years. The cropping system analysis takes into account the influence of factors like soil and weather, the influence of some other factors are omitted or considered constant.

Water Management Module

This module is responsible for the computation of monthly and annual water balance; this water balance is the difference between potential evapotranspiration and rainfall. The water resource of Punjab is estimated as the sum of surface and ground water; a water resource map will be produced. For the monitoring of the ground water level and quality, different maps based sample data from wells and piezometers will also be developed. For the day-to-day and long-term management of the water resources, an ensemble of models will be integrated to interact with the AGCN core system. These models include ground water model, hydrologic and flood forecasting models, reservoir model, hydraulic and flood mapping models. This module is developed in a companion paper entitled "Implementation of an Integrated Decision Support System for Water Management in Souss-Massa Morocco".

AGCN SYSTEM ARCHITECTURE

There are five parts of the overall AGCN system from a high-level perspective:

1. Agriculture Data Input includes communication, digitization and entry of relevant agriculture data (from

weather to water, soil and statistics). The source of data could be either digitized disk file data or on-line/in-situ data transmitted through the network.

2. Application-Specific Services are high level services that are specifically responsible for handling application-dependent data access and logics/models. Examples are from Weather Services to Water Services to Yield Forecast Models.
3. Core Geo-Capacity Center Services and Data Management provides the core generic spatial (raster and vector) data processing services, spatial data access services, and spatial data storage and management.
4. Applications are specific agriculture applications built on top of both the Application-Specific Services and the Geo-Capacity Center. Examples are from Wheat Yield Forecast to Water Table Monitoring and Post-Harvest Planning/Management.
5. Products are results generated by the Applications. They could be either maps or reports or both.

To build the entire AGCN system, enterprise system architecture will be used. Enterprise system architecture meets the main system requirements as data centralization for efficient access and management, data distribution for sharing and storage efficiency, services oriented architecture for best-of-breed technologies integration, multi-tier architecture for efficient & effective system enhancement, and componentization environment for system extensibility and scalability. The available enterprise technology and relevant information technology (IT) are matured enough for implementation of the AGCN system and also ensure the security and performance requirements.

The inclusion of the Geo-Capacity Center in the AGCN system turns the enterprise system into a Geo-Capacity Information System (GCIS), which provides the ability to handle spatial raster/vector data. The spatial-enabled ability allows the AGCN system to practicably solve real-world agriculture problems, such as spatial-based monitoring, forecasting and management. The main technologies required for implementing the Core Geo-Capacity Center Services and Data Management are as follows:

1. Image processing technology. Abundant libraries of various image processing algorithms exist nowadays. The maturity of these algorithms has been proved by many applications for decades.
2. Componentization environment for the image processing technology. Recent technology advancement in pluggable function framework provides a componentization environment for applying mature processing algorithms.
3. Application Programming Interface (API). Application Programming Interface and Software Development Kit provide an environment for users to develop their own algorithms in the pluggable function framework which an off-the-shelf system does not provide. The new pluggable functions developed can be easily deployed into the componentization environment.
4. Spatial Database. Spatial database provides an efficient data structure (objectrelational table) to store spatial raster/vector data and optimize query performance.
5. Geo-Database Management (GDM) technology. GDM technology handles different formats of raster/vector data and works seamlessly with spatial databases.
6. OGC Web Services. OGC (Open Geospatial Consortium) provides international standards for publishing raster and vector data for sharing among different data providers. These standards and technology provide relevant products and services (Web Feature/Map/ Coverage Services) for data publication, query and retrieval.
7. RAID Storage System. The current RAID hardware storage system can hold huge amounts of data, up to Terabytes of data, to meet the data storage requirement for the AGCN system. More importantly, the RAID system provides maximum reading/writing speed and reliable data safety for the AGCN system.

To develop high-level Application-Specific Services for the AGCN system, Application programming interface and software development kit technology can be used to develop the corresponding pluggable functions and componentization environment required by the AGCN system.

The development of the specific Applications could be diversified using different tools. Maintenance of a standard communication protocol (such as HTTP and/or SOAP) between the specific applications and the backend services (including the core Geo-Capacity services) is required.

The Products are usually presented in a map format. The core Geo-Capacity services provides components/tools for data analysis and map generation.

The application/technical view covers the major components and their inter-connection in a multi-tier enterprise framework for the applications:

1. Data tier. This tier is responsible for storing and managing the agriculturerelated spatial data. A spatial database is recommended. Multiple spatial database servers can work as a grid system or as remotely distributed systems.
2. Services tier. This is responsible for providing services to access and process data, conduct agriculture-relevant analysis, and disseminate user-required information. It can be divided into two sub-tiers:
3. Data access sub-tier. This tier will provide a standard interface to access data from either homogeneous or heterogeneous database systems. Upon availability, the data can be registered and published for access from permitted services.
4. Services sub-tier. This tier consists of various agriculture-related components to provide services from data processing to monitoring, prediction and management.
5. Presentation tier. This tier contains agriculture specific applications which integrate the available services and provide user interfaces for their clients.

CONCLUSION

The AGCN for the implementation of an Agriculture Information system is based on the Geo-Capacity Information System (GCIS) concept and is conceived with the aim of developing methods to produce timely statistics on land use, planted area and production volumes for various crops within India, of applying remote sensing and ground surveys to estimate the planted area, as well as providing weather information and water management service applications and tools. Real-time image processing tools, proven methods to relate satellite imagery to quantitative crop yields, weather, soil and crop information to feed crop growth monitoring and water models, information technology required to build an open and multi-tier architecture are presently available, such that implementation and deployment of an AGCN system is not beyond our reach.

2

Pest Detection in Cropping Systems Through Remote Sensing

The major cropping systems are rice-wheat, rice-mustard-rice, cotton-wheat, pearl millet-wheat, rice-sugarcane and maize-wheat in Indo-Gangetic Plains of India. The Indo Gangetic plains regions are divide in four parts viz. Trans Gangetic Plain, Upper Gangetic Plain region, Middle Gangetic Plain and Lower Gangetic Plain region. The crops are rice, wheat, maize, cotton, sugarcane and pearl millet. These crops are grown in large areas of IGP in rabi and kharif seasons.

The cropping systems are largely affected due to pest. The major pests viz. aphid, stem fly, seed fly, Thrips, Shoot fly (Atherigona soccata), pod borer, moth and ball worms are affected seasonal crops. Due to affect of pest, adverse affected on the crop production and yield. Remote sensing technology plays a great role to prevent and control and management of pest. The insect affected crop and normal crops are give different tonal variation in imageries. The LISS III and LISS IV imageries are very helpful for detection of affected area. The affected areas are identified with the help of GPS (LEICA Differential GPS). The GPS give exact location and position on the imagery. The image interpretation keys shape, size, tone, texture and association help to detection to visual on the imagery and confirmation during ground survey. After image analysis found the great variability in tonal variation on the imagery.

Normal crops give red, bright red and dark red with smooth texture but pest affected area give pink, yellow and yellow pinkish red colour with irregular shape and rough texture. Remote sensing technology help to season wise analysis of pest in affected area and help to control through crop rotation, if possible and through chemically.

INTRODUCTION

The different types of information on natural resources such as soil survey information, land use, vegetation information etc. can be extracted from the same satellite data. Remote sensing as a tool will give information regularly and at a low cost to enable timely reclamation. In India, production forecasting of certain crops, crop yield modeling and crop stress detection are done using remote sensing data. The first satellite remote sensed data became available in 1972. Therefore, we had the opportunity to get an overall synoptic view only after 1972. Currently a number of spacecraft imaging systems are operating using remote sensing sensors. Some of the current image systems from spacecraft platform include Indian remote sensing satellite (IRS), SPOT and IKONOS etc.

The variation between signatures, when the bands are assigned the primarily colours of red, blue and green, one can visualize the image as a colour one. Vegetation appears red in such a colour composite where NIR band is assigned as red, due to high reflectivity of vegetation in this band. This is typically termed as False Colour Composite (FCC).

The temporal IRS data were screened to select cloud free data during winter and summer season. Care was taken to select data pertaining to the period when crop coverage was with full foliage. The villages appear bright while water bodies had low backscatter and appeared in dark black colour. This was used to map the villages and water bodies using a decision rule classifier. All these two classes: villages and water bodies were combined and made a mask. The rest of the area was considered as potential agricultural area.

Detection and identification of plant diseases and planning effective control measures are important to sustain crop production. One of the potential applications of remote sensing technique in agriculture is the detection of plant disease an extensive areas before the symptoms clearly appear on the plant leaves

This has been clearly demonstrated for the first time in remote sensing studies on root wilt disease on coconut plans in Kerala under a NASA-ISRO-IARI collaborative programme (Dakshinamurti, 1969 and Dakshinamurti et al., 1970). Healthy plants give a higher reflectance in the near-infrared region and a lower one in the visible region. Diseased plants show a higher reflectance in the visible spectrum and a lower one in infrared region. This principle can be used in distinguishing healthy and diseased vegetation and assessing vegetation damage due to diseases.

Crop intensification is a viable proposition as far as India is concerned as it is endowed with favorable climatic condition for growing a diverse range of crops practically through out the year (Panigrahy and Manjunath, 2003). This concept of looking to the agricultural land as a system is known as the cropping systems approach that seeks to sustain the increase in the food production by judicious combination of crop intensification and yield improvement. Cropping pattern is defined as the yearly sequence and spatial arrangement of crops or of crops and fallow on a given area. The cropping pattern is categorized on the basis of crop season as Kharif, rabi and summer season cropping pattern. Growing two, three and multiple cropping patterns. The productive base of cropping system is plant growth, which is influenced by management and environment. Management here includes all those components associated with crop production. These include, crop area, crop biomass, crop rotation crop calendar, crop biomass, time and spread of sowing and harvest. Environment is the invariant resource of soil, physiography and climate.

STUDY AREA

IGP region divided in four agro climatic regions TGP (Trans Gangetic Plain), UGP (Upper Gangetic Plain), MGP (Middle Gangetic Plain) and LGP (Lower Gangetic Plain). These four zones are divided in 16th sub zone. The Indo Gangetic Plain region of India has total geographical area of about 47.4 million hectares (Pandey et al. 2006). The rainfall ranges from high as 3000 to low 300 mm. Administratively, the states are divided out of 203 but only 173 districts are belong to IGP and encompasses the entire states of Punjab, Haryana, Uttar Pradesh, Bihar and West Bengal. The major rivers of IGP are Ganga, Yamuna, Satluj and Ghagra. Soils of this region are loamy to sandy loam. The percentage of the organic carbon is enough to other region. The study was confined to Indo Gangetic Plain Region. These four agro climatic regions as defined by MOA (1997) and Agro-climatic Zones of India Planning Commission Report (2005) were further classified into 16 different sub regions. Three cropping seasons namely viz. Kharif (June to September), Rabi (October to February) and Zaid (from March to May) were taken into consideration.

MATERIAL AND METHODOLOGY

i. IRS IC /1D LISS III (FCC) digital data
ii. District maps and state maps
iii. Indian Agro climatic zone map
iv. Global Positioning system (TOPCON and GARMIN) and
v. Performa for field data information

These data and instruments were used in the study.

The methodology consists of the study and analysis of remote sensing data, collection of ground truth information, collection of cropped and pests of field data and creation of data base for all the required physical and cultural parameters and overlying, integration and analysis of all spatial, non spatial and attribute data bases.

Methodology

Analysis of remotely – sensed data in different forms and formats was analyzed after examining the suitability of various products. The study proceeded as follows:

1. Preliminary study of the IRS IC /1D LISS III – FCC.
2. Study of false colour composites on digital (onscreen) or paper prints
3. Preparation of image interpretation keys

4. Collection of ground truth with GPS and
5. Preparation final output map

The digital remote sensing data were processed and analyzed using the ERDAS IMAGINE 8.6 software. Initially, the digital data of scenes were geo-referenced. The Kharif and rabi season images were considered as the base since it is very sharp and clear. This image is first geo referenced by taking various control points from the SOI toposhits. The projection type used is 'polyconic' with the spheroid and datum as 'modified Everest'. Subsequently, image-to-image registration was performed in order to register the second image and so on. Some sharp and easily identified features like crossing of roads, railways, canals, bridges etc. were located on both images and were selected as ground control points. Some GCP's with rms error beyond threshold were deleted and replaced by other points so as to maintain acceptable positional accuracy.

Preparation of Questionnaires

The necessity for acquiring such as information was felt on the basis of secondary information and the type of information acquired by direct observation during the reconnaissance field visit.

Preparation of Base Map

The base maps indicating permanent and artificial features such as canal, highway, roads, railway, irrigation network, and districts boundary were prepared in ARC GIS 8.3 software. For this, Survey of India sheets were used as reference materials. The permanent fixtures were transferred in digital format through digitization.

Survey of the Area/Reconnaissance

Surveyed of the area season wise in a year viz. kharif (June), rabi (January) and zaid (April). Reconnaissance of the area of IGP's states, collected normal and pest affected crop samples and related information along with GPS points in different fields. Global Positioning System burgeoning technology, which provides unequalled accuracy and flexibility of positioning for navigation surveying and GIS data capture.

The GPS uses satellites and computers to compute position any where on the earth. GPS viz. TOPCON, GARMIN and Differential GPS are very good for agricultural purpose.

Collection of Ground Truth Information

Field visits were made to establish the relationship between the image elements, namely colour, texture, pattern, shape, size, association, etc. and the existence of normal crop and pest affected crop. Initially, reconnaissance traverse of the area was made to assess the trafficability and to precisely located sample areas, the pockets of land which were interpreted as normal as well as pest affected crop were precisely located on the ground with the help of topographical sheet and observations with respect to terrain conditions. Observations were, however, also made outside the sample areas randomly, in order to validate the relationship already established between image elements, and normal and pest affected crop areas. During surveyed collected crops and pests information 200 points with geo coordinate information these information overlaid on the imagery and observed visual as well as digital.

GIS AND REMOTE SENSING ANALYSIS FOR CROPPING SYSTEMS

GIS Analysis for Cropping Systems

The districts polygon coverage were digitized for the IGP region in India covering the states of Punjab, Haryana, Uttar Pradesh, Bihar, West Bengal, and a part of Rajasthan in ARC GIS system. A database was developed for analysis of cropping systems in MS Excel, converted dBASE IV format then finally converted in .shp file to overlay analysis in ARC GIS as well as ERDAS IMAGINE. Agro climatic zone, sub zone, districts and states information were taken from planning commission report. The coverage has been generalized based on the yield in to two classes, i.e. Normal crop and Pest-affected crop depending upon the range of crop survey data.

Remote Sensing Analysis for Cropping Systems

The post fieldwork consisted of modification of preliminary interpreted mapping units on FCC of satellite data in the light of field information data. The tentative legends prepared also finalized. The normal crop and pest affected crop areas; training sets were finalized on the computer screen with the help of GPS (latitude/longitude) points.

Visual Analysis of FCC's

The tonal variation and with the help of image interpretation keys identified the normal and pest effected crop area. During the image interpretation different crops have different tonal variation so it is very essential to study very precise and have to keep mind pre information about the area. The pest-affected areas are given the different tonal variation in the normal crop area. The normal crop give red, bright red and dark red colour with smooth texture in regular form but pest affected areas give pale yellow, yellowish red and dull red or pink colour, these tonal variation mixed with other crops. The tonal variation varied crop to crop in different season and in different zone. Our concentrates were detecting the pest-affected area identified after observation.

Digital Image Enhancement

Enhancement is the modifications of an image alter its impact on the viewer. This improves the visual interpretability of an image by increasing the apparent distinction between the features in the screen. Histogram equalization is one of the techniques of non-linear contrast enhancement. Histogram equalization (uniform Distribution Stretch) provided maximum contrast for identifies normal and pest affected area. Filter algorithms for performs image enhancement, they suppresses certain frequencies (de-emphasize) and pass (emphasize) other frequencies. In the present study the following image enhancement were performed viz. linear enhancement and non-linear enhancement etc.

Image Analysis

The geo-referenced digital data of IRS LISS III sensor for the two different seasons (Kharif and Rabi) were used to prepare the land use/ land cover maps for IGP. The digital image processing techniques used for land use/ land cover classification includes (a) image enhancement (b) image classification. Digital analyses for land use/ land cover classification were carried out using the image processing software-ERDAS IMAGINE 8.6. Based on ground truth data, training sets are identified for various land use/ cover classes. Using these training sets, the land use/ land cover classes were delineated employing maximum likelihood classifier in ERDAS IMAGINE. The thematic maps of land use/cover for both the scenes as well as the pest-affected crops were merged to prepare the land use/ cover maps of IGP. Various land use/ cover classes delineated are built-up area, villages, water bodies (ponds), road, forest, railway line, and river and sandy area.

The tonal variation and with the help of image interpretation keys identified the normal and pest effected crop area. During the image interpretation different crops have different tonal variation so it is very essential to study very precise and have to keep mind pre information about the area. After this process prepared the digital classifications, run the un-supervised (ISODATA) and supervised classification (Maximum likelihood classification).

After run the supervised classification two broad classes are observed (1) normal crop and (2) pest affected crop areas.

Normal cropped area: In Kharif sesason's imageries identified major crops are rice, maize, bajra, jowar, Pigeonpea, green gram, black gram, groundnut, sesamum, lahi, and berseem. In rabi sesason's imageries major crops are wheat, sunflower, barley, potato, chickpea, pea, mustard, oat and sesamum identified. The normal crop gives red, bright red and dark red tone with smooth texture in regular form

After analysis IRS imageries observed season wise like that in Kharif season rice give dark red colour, sugarcane give bright red colour, in rabi season wheat give red colour pulse give red pink colour so different types give red tonal variation on the imageries.

Pest affected cropped area: Pests are affected in both seasons (rabi and Kharif) on the different crops. The major pests are in crops (1) rice-hispa, gundhi bugs and jassids (2) cotton-boll worms, leaf roller and whitefly (3) maize-grass hopper and thrip (4) pearl millet-midge fly and termite (5) sugarcane-top borer, pyrila and mealy bug (6) chickpea-pod borer, bruchids and cutworm (6) jute-leaf beetle and caterpillar affected in Kharif season.

The major pests are in crops (1) mustard-aphid and painted bug (2) pigeon pea-pod bug and pod fly (3) barley-stem borer and grasshopper (4) wheat-termite, thrip, and gujhia weevil (6) sunflower-capitulum borer (7) berseem-black ants (8) potato-aphid, white grub and tuber moth affected in rabi season. The pest affected areas give pale yellow, yellowish red and dull red or pink colour, these tonal variation mixed with other crops.

Major cropping systems in IGP: The cropping systems livelihood are based but pest affected the crops and diseased yield and production after survey analysis of data zone and sub zone wise the dominant cropping system observed.

The major cropping systems were observed after data analysis of zone wise information of the IGP. The zone wise cropping systems are in (1) TGP-rice-wheat, cotton-wheat, maize wheat and cotton berseem (2) UGP-rice-potato, maize-potato, maize-wheat and pigeon pea-wheat (3) MGP-rice-lentils, rice potato and jute –rice-wheat and (4) LGP-rice –rice-mustard, rice-potato and rice-wheat-jute.

In the IGP, the maximum area was covered by rice – wheat-cropping system. In the Lower Gangetic Plain, rice-wheat cropping systems have lesser area because of boro rice is major crop. The cropping system is also changing in TGP, the south-west Punjab areas (Bhatinda and Muktsar districts) cotton belt converting in to rice belt. These changing are also due to cotton bollworm problem. Haryana and to some extent in western Utter Pradesh owing to high water table during monsoon season, rice has become a major crops in such areas.

Remote sensing technique is very useful, less time consuming and cost effective in delineation of pest-affected areas. The GPS, DGPS, LAPTOP and other sophisticated instruments give the very precise information and remote sensing software ERDAS IMAGINE play a vital role and reliable information to making pest control program and identified the pest affected area of the IGP. Immediate attention needs to be paid for control the pest affected area and prevent affected areas through chemical. This would facilitate enhancing the area under rabi and Kharif cultivation and increasing agricultural production, for better livelihood of the people.

The pest largely affected crop area and control is essential before the effect so remote sensing technology play a great role to control pest affect the area. It gives very good information before if precisely identified the cyclic study the remote sensing data of the area. The interchange the cropping systems also help to control the pest problems. The major pests are observed after data analysis, in kharif season-gundhi bug, hispa, jassids, bollworm, grass hopper, thrips, top borer, pyrilla, aphids and white grub; in rabi season –termite, cutworm, whitefly, tuber moth, hairy caterpillar, hooded hopper, hawk moth, painted bug and gall fly (Singh et al. 2003).

IRS 1C/D, LISS III images given the area information very accurately. After visual interpretation run different algorithm and digital classification these data given valuable information of this area. Most of the growing cropped area some field are badly affected due to pest. Interpretation keys play very important role in the visual interpretation of image for identification of normal and pest affected areas. Due to pests, crops have poor growth and poor yield.

Fire blight is the most dangerous disease on apple in the north-west of Iran (Azerbaijan province).Weather-based fire blight risk maps of the east Azerbainan province were made by Geographic Information System (GIS) with the use of 20 years data from 10 weather stations using the MARYBLYT model. At first step, the long term meteorological data was assess using STATISTICA software for calculating conditional probability of occurring a day with suitable thermal and humiity conditions which fire Erwinia need for flower infection and outbreak in the province. Then, by using Geographical Information System (GIS) and preparing the digital elevation model (DEM) of province the map of fire blight risk was prepared. The results of this study show that regions of west and north-west of province are of high fire blight risk.

Geographic Information Systems/Science (GIS) seem to be a natural fit for epidemiology, the study of the origin and transmission of disease. Epidemiology attempts to integrate a vast array of risk factors such as temperature, moisture, vegetation type/percent cover, the presence and density of disease vectors (i.e. mosquitoes, ticks, etc.), the presence and density of susceptible hosts, pathogen transmission rates-even the molecular, cellular, reproductive and behavioral biology of the host(s)-in an orderly way. This information is used to identify which factors are necessary for disease, the areas that support disease vectors, and the risk of disease transmission and outbreak. A GIS provides us with the ability to overlay many of the larger-scale risk factors and spatially analyze the data in an attempt to better understand which factors are essential, or how the factors influence each other, to increase or decrease the occurrence of a particular disease. The articles summarized below provide an introduction to some epidemiological studies which have utilized GIS technology Fire blight, caused by the bacterium Erwinia amylovora, affects over 130 plant species in the rose family.

In the east Azerbaijan province Fire blight is a destructive bacterial disease of apples and pears that kills blossoms, shoots, limbs, and, sometimes, entire trees. Once the first few open blossoms are colonized by the bacteria, pollinating insects rapidly move the pathogen to other flowers, initiating more blossom blight. These colonized flowers are subject to infection within minutes after any wetting event caused by rain or heavy dew when the average daily temperatures are equal to or greater than 60 F (16 °C) while the flower petals are intact. Although mature shoot and limb tissues are generally resistant to infection by E. amylovora, injuries caused by hail, late frosts of 28 F (-2 °C) or lower, and high winds that damage the foliage can create a trauma blight situation in which the normal defense mechanisms in mature tissues are breached and infections occur (Infected flowers first appear water-soaked, then shrivel, turning brown or black. As the infection progresses, leaves on the same spur turn dark brown or black as thought scorched by fire. The dark, shriveled leaves hang downward and usually cling to blighted twigs. Infected shoots, twigs, and suckers turn brown to black and often bend in a characteristic shepherd's-crook. Infected immature fruit turns dark, shrivels, mummifies, and rots. Mummified fruit may cling to the tree for several months. A canker is formed when an infection progresses into larger branches

Climatic Factors and Fire Blight

A very important aspect of fire blight management involves monitoring of weather for the specific conditions that govern the build-up of inoculum in the orchard, the blossom infection process and the appearance of symptoms. A weather station record of daily minimum and maximum temperatures and rainfall amounts is needed. When 50 percent of the buds show green tissues, begin keeping a daily record of the cumulative degree days (DD) greater than 55'F (12.7 °C).This information can be used to signal when symptoms are likely to appear in the orchard for blossom blight [103 DD greater than 55 F (57 DD greater than 12.7 °C) after infection], canker blight [about 300 DD greater than 55 F (167 DD greater than 12.70 °C) after green tip], and early shoot blight [about 103 DD greater than 55 F (57 DD greater than 12.7 °C) after blossom blight or canker blight symptoms appear]

Temperatures

As the blight bacteria can thrive only about four days on each individual flower, the bacteria must develop to dangerous numbers during the immediate three or four days leading up to blossom wetting. Warmer temperatures allow rapid bacterial growth in flowers. If bacteria numbers exceed a certain minimum while the flower is in good condition, then the flower is lightly wetted, infection is possible.

Blossom Wetting

Blossom wetting alone does not cause fire blight. Rain during cold or cool weather does not lead to infection, or blight would be common everywhere, every year. However, if recent temperatures have allowed rapid bacterial development, blossom wetting is the infection "trigger." Moisture can be from rain, dew or light irrigation wetting. The water moves the bacteria from the stigma tip into the flower's nectary. The bacteria may then grow rapidly, attack the fruitlet, and then move into the tree, attacking the living portion between the wood and the bark.

Fire Blight Risk Assessment Models

ClaimsFire blight risk assessment models are briefly discussed below:

Maryblyt Model: The most successful computerised prediction system developed in the United States is the MARYBLYT™ model. MARYBLYT™ has been trialed for over 10 years in the United States and is now being used by American growers and consultants to predict fire blight. The MARYBLYT™ programme uses the infection criteria to identify infection events and predict the development of symptoms. These criteria to identify infection events are:

- Flowers open with stigmas and petals intact
- The accumulation of at least 110 degree hours when temperatures are greater than 18.3°C after first bloom
- Wetting caused either by dew or 0.25 mm of rain (or at least 2.5 mm of rain the previous day).
- An average daily temperature of at least 15.6°C.

Computer printouts give the daily temperatures, wetness and likelihood of infection which can be interpreted to decide whether a streptomycin spray is required.

Cougar blight model: During the flowering period and on the basis of degree hours accumulation, this model can forecast the risk of infection of Ervina in host tree.The minimum four-day Celsius degree hour accumulation that appeared to be necessary for infection to occur during a blossom wetting was set at approximately 270 °C (500 F). The degree hours can be calculated with bellow equation:

DH=Sum(Ti-Tb)

Where:

Ti:hourly temperature

Tb:base temperature

DH:degree hour

Further experience has shown that this number should be considered a usually trustworthy guideline only, and the degree of infection risk as determined by degree hour accumulations should be considered as an ever-increasing curve upwards, with "no risk" at the base, rising through "possible, but unlikely, infection" to "probable infection," "high risk of infection" and "extreme risk of infection." Stair-step thresholds were found to be an improbable model assumption for two reasons: first, The ultimate size of the colony after a specific amount of time and heat may be greatly influenced by its' CFU(what is cfu) numbers when it was first placed upon the stigma surface, and second, It is not likely that a single degree hour unit would be sufficient to actually move risk from "moderate" 269 four-day degree hour total to a "high" 270 four-day degree hour total.

Only the days that trees are blooming are counted. Daily degree hours are forecasted for the three days in future using the look-up chart and their most trusted forecasting source. The value for each of the days is added to the sum of the days that precede it. Each morning, the user can adjust the model, updating the previous days forecast to an actual high and low temperature degree day value, sum the current four previous days total and a new three-day forecast.

After the third day of blooming, each present day degree hour value is the sum of the past three days' degree hours, plus the present days forecasted value. The grower can judge the daily fire blight risk from the present four-day degree hour total. As the four-day degree hour total rises to the level that indicates "extreme" risk, growers are advised to apply all available preventative materials on their recommended schedule until the risk drops, or the blossoms are no longer on the trees. In this study based on climatic conditions of east province we found the high risk areas to fire blight outbreak.

MATERIAL AND METHODS

Studied Area

Figure 1 one shows the geographical location of the East Azerbaijan province in Iran. with an area of 45480 Km2 East Azerbaijan is one of the northwest provinces of Iran. Daily mean,maximum and minimum temperature, daily relative humidity data as well as precipitation information were obtained from the synoptic and climatic stations of the area for a 25 year period (1980-2004) (Table 1).The phonological (flowering date) data of apple and pear also were collected from ministry of Agriculture for the same period.

Fig1-Geographical location of east Azerbaijan in Iran

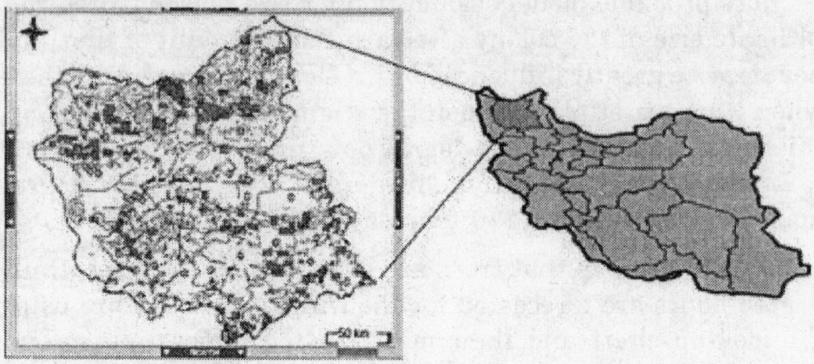

Tab.(1): List of meteorological station of area

Num.	Station	Elevation(m)	Latitude	Longititude
1	Sarab	1651	37 56	47 32
2	Bostanabaf	1750	37 50	46 50
3	Mianeh	1094	37 25	47 42
4	Tabriz	1361	38 05	47 16
5	Maragheh	1420	37 24	46 16
6	Sharafkhane	1302	38 11	45 29
7	Khodaafarin	300	38 08	46 56
8	Azarshahr	1400	38 26	45 45
9	Jolfa	704	38 57	45 38
10	Ahar	1300	38 29	47 03

Information Layers Preparation in GIS

In this study based on the MARYBLYT model we prepared the temperature and relative humidity maps of area and then by using GIS we found the area of the province that their climatic conditions were suitable for spreading and infecting the orchards.

Digital elevation model(DEM): Digital elevation model(DEM) was prepared using topographic map. For this purpose Arc/Info software and a Digitizer were used to digitize all necessary features presented on the map based on UTM map projection system. By exporting the layer prepared in Arc/ Info to IDRISI software the format of the map was converted from vector to raster

Temperature Map: The phonological data of apple and pear shown that the flowering period of these two host in east Azerbaijan is during April and may. so,we first calculated the probability of occurrence of daily temperature between 18.3 and 30.0°C for all meteorological stations available in the area.To do this we used daily temperature data and HYFA software to calculate probabilities of occurrence of daily temperature between 18.3 and 30.0°C in during flowering date of Apple. Then the probability map of occurrence daily temperature between 18.3 and 30.0°C during the flowering period was prepared using IDRISI software.(figure2).

Fig. 2)Probability map of a daily temperature between 18.2 and 30 degree centigrade during the flowering of Apple

Probality map of occurring a wet day with suitable temperature for fire blight: Wetting period is one of the most important factors affecting the epidemy of disease. So, the daily relative humidity and rainfall amount of all meteorological stations available in the province was first analysed.. For each station, the probability of occurrence of daily relative humidity greater than 60% or rainfall amount greater than 0.25 mm was calculated.. To do this, we extracted in each station days which had daily relative humidity greater than 60% or rainfall greater than 0.25 during the flowering date of host tree from the historical data were extracted. Finally the probability map of wetting days(RH>60% or rainfall>0.25) during the flowering date of apple and pear were prepared using Statistica and IDRISI software.(fig. 3)

Fig 3-the probability map of wetting days(RH>60% or rainfall>0.25) during the flowering period

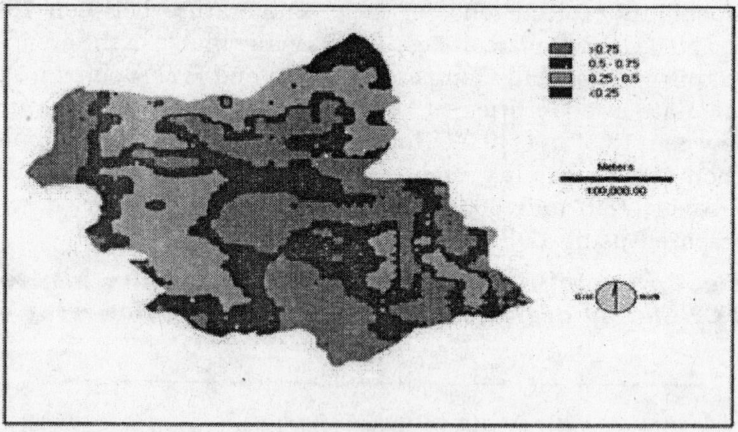

RESULTS AND DISCUSSIONS

For preparing of the fire blight risk map the conditional probability of occurrence of a day with daily temperature between 18.3 and 30 degree centigrade and relative humidity greater than 60%(or rainfall>0.25) should be calculated. To do two probability map of temperature(figure2) and moisture(figure3)were combined using the Analytic Hierarchy Process (AHP).

The result map contain the probability map of suitable condition for infection and outbreak of fire blight during the flowering period. In this map, fire blight risk areas were determined according to the climatic parameters. Finally we reclassified the province to four fire blight risk class: low risk, medium risk, high and very high risl level. (figure4).

Fig. 4-fire blight risk map of the east Azerbainan province

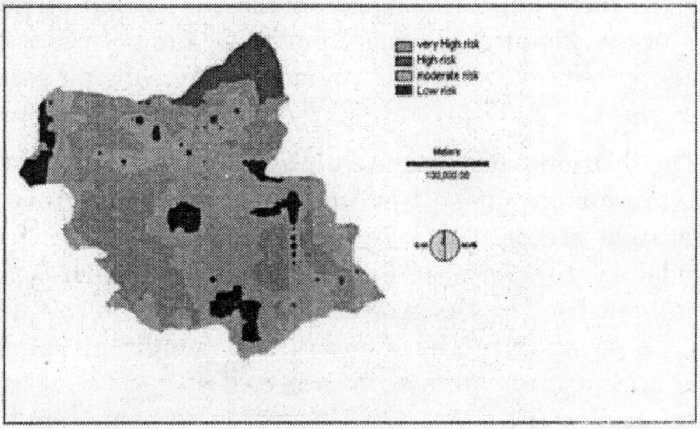

Our investigation shows that the risk of fire blight epidemy is very high in the west and northwest of Azerbaijan province because of the presence of climatic factors which affect infection and outbreak of fire blight. In this regions, the probability of a wet day with daily temperature between 18.2 and 30.0 degree centigrade during the flowering period is very high. This probability was calculated in this research according to the historical daily temperature, rainfall and relative humidity of meteorological stations available in the province. In the Mountainous area of Azerbaijan province with elevation more than 2500 m have a not unsuitable climatic condition for fire blight. The other areas of province such as Tabriz, Azarshahr and Jolfa have the medium and low risk.

A GIS AND REMOTE SENSING BASED APPROACH TO DEVELOP COLD STORAGE INFRASTRUCTURE FOR HORTICULTURAL CROPS

Horticultural crops cover a large number of fruits, vegetables, flowers which are highly perishable in nature. Post harvest losses estimated to be in the range of 20-40 per cent. Hence, emphasis is given to develop post harvest infra-structure like cold storage, food processing, packaging, market outlets etc. during the current plan period. Most of the cold storages are concentrated in and around the consuming markets.

Thus, very little facility exists to cater to the marginal farmer's requirement during the harvesting season. Potato is one of the most important vegetable crop. Though 90 per cent cold storage facilities of the country is for potato crop and located in the potato growing regions, still it falls far below the requirement.

The National Co-operative Development Corporation is trying to promote the setting up of such storage facility in the co-operative sector.. Thus, a scientific approach to evolve a methodology to locate sites for cold stores which would be optimally utilised by the growers is required. A pilot study was taken up to analyse the demand and supply situation and evolve an optimum plan to locate cold stores using satellite remote sensing (RS) data and Geographic Information System (GIS). The study was done for potato crop in Bardhman district of West Bengal, a leading potato growing area.

MATERIALS AND METHODS

Study Area

Bardhaman district is one of the major potato growing districts of West Bengal. The study was conducted for three police stations of Bardhaman district, namely, Jamalpur, Kalna and Memari which account for 82.9 and 83.3 per cents of potato area and production of the district respectively. There are 35 potato cold storages in the study area.

Data Used

Satellite based RS data of IRS 1C WiFS (188 m resolution) was used for potato crop map generation. For getting the road network and settlement location high resolution RS data of IRS 1C LISS-III (23 m resolution) and Survey of India toposheets were used. The information about the cold storage locations, cold storage statistics (capacity, ownership, area of jurisdiction) and potato crop statistics were collected from Department of Agricultural Marketing, West Bengal.

Crop Map Generation

The RS data was georeferenced and the boundary mask of Bardhaman district was overlaid. The extracted data was classified using a Maximum Likelihood classifier. Field information collected synchronous to the satellite pass was use for the supervised classification.

GIS Analysis

Spatial database was made for all the spatial data, like, road network, settlement locations, cold storage locations using Arc/Info GIS software. The database of the statistics of the cold storages was also linked tc the spatial data. The raster image of the crop map, generated from satellite data, was converted to points coverage. Two steps of analysis was used for locating optimum cold storage sites. Those are buffering analysis and location-allocation analysis. In buffering analysis buffers were created around villages and already existing cold storage locations. The buffer sizes for the cold storages varied as per the capacity of the storage. Village buffer sizes were also dependent upon the village sizes. A large number (96) of probable cold storage sites were marked using the following criteria :

i. the cold storage should be near a major road,

ii. it should be within the buffer of a settlement,

iii. it should not be within the buffer of the already existing cold storages.

Using the already existing cold-storages and the new probable sites location-allocation analysis was carried out to a) locate the required number of optimum new sites and b) allocate crop points to cold storage.

Results and Discussion

i. Analysis of Data for Existing Cold Storage: A database of the information collected regarding all the 35 cold storages in the study area were analysed using Dbase. The analysis showed that the existing cold storages have a total capacity of 281.3 thousand tonnes which is remarkably low compared to the total production of potato (738.9 thousand tonnes) the area (Table 1). Assuming that ten per cent of potato will be used for local consumption, there is still a storage infrastructure deficit of 383.8 thousand tonnes. The average storage capacity of the existing cold storages is 6.74 thousand tonne. Hence there is a need of at least 57 new cold storages, of the average capacity, to store the additional potato production.

ii. Potato Crop Map Generation

IRS WiFS data was digitally classified using a Maximum Likelihood classifier with the help of field information. The overall classification accuracy achieved was 92.0 per cent (Panigrahy and Chakraborty, 1998). The high accuracy achieved with the medium resolution (188 m) data was due to the fact that the potato in the study area is mostly grown in concentrated area, especially near rivers. The total estimated potato acreage of the study area was 28.7 thousand ha. A potato crop map was generated by masking all other land cover classes. The pixels in the map were converted to points in the Arc/Info coverage for being used as demand points in locating and allocating cold storages.

iii. Buffering Analysis

Buffers are the zones around the geographic features. The buffers were generated around the settlements (big villages and townships) and cold storages. The buffer radius for each cold storage has been calculated based on its capacity. A buffer of 111 m was created for 1,000 tonne capacity of a cold storage. This buffer size was decided by taking the Bardhmann district average yield value of 257.06 quintals per ha. The settlement buffer was of the radius of 500 m except for three large townships (Memari, Kalna and Jamalpur) where it was 2.0 km. These buffers were used for marking, interactively, a large number of probable new cold storage sites (96 in number) as per the criteria set in the previous section.

iv. Location-Allocation Analysis

Minimum distance criteria was used in the location-allocation analysis to optimally locate the new cold storage sites from the probable sites and allocate the crop points to the cold storage locations. Since the crop points are generally not connected by road a radial distance method has been used. The allocation showed that 5 out 35 existing cold storage are so injudiciously located that not a single crop point is allocated to them. Even when the analysis was repeated for only those 35 existing locations 4 did not get any allocation. The average distance to travel from crop point to cold storage reduced from 3.1 km to 1.6 km by incorporating the 57 new sites (Table 2). The maximum distance of travel also decreased from 5.6 to 3.6 km. While the average load on the existing storages were 210.7 thousand quintals it would be decreased to 79.0 thousand quintal on by incorporating these new locations.

All the new suggested cold storages have been classified into three groups based on the number of demands allocated to them. The first 11 sites are of highest priority are those sites which are catering more than 200 crop points. There is an urgent need to create infrastructure for new cold storage in these sites. The 5 original cold storage sites which are not getting allocated any crop site can also be relocated to the new sites suggested. The study shows that remote sensing and GIS can be used to develop a scientific approach to draw an integrated plan for cold storage, food processing, packaging sites etc. for fruit and vegetable crops.

Table 1.Statistics of potato crop and cold storage in three police stations as compared to Bardhman district.

Parameter	Bardhmann	Study Area
Acreage ('000 ha)	34.6	28.7
Production ('000 t)	886.6	738.9
Cold Storage Number	59	35
Cold Storage Capacity ('000 t)	355.1	281.3

Table 2. Improvement achieved in potato cold storage allocation

Parameter	Existing Cold storage	Suggested Cold Storage (Existing + New)
Total No. Of Cold Storage	35	92 (35+57)
Travel Distance (Crop site to Cold Storage), km		
Maximum	5.6	3.6
Minimum	607	1172
Average	3.1	1.6
Capacity Required (‘000 q)*		
Maximum	614.0	367.0
Minimum	0.91	0.91
Average	210.7	79.0

*Average capacity of existing cold storage is 67.4 thousand quintal.

OPTIMUM SYNTHETIC APERTURE RADAR SYSTEM PARAMETERS FOR RICE AND TROPIC VEGETATION MONITORING

Rice is the staple grain in southeastern Asia. Its monitoring and yield estimation has special significance to this region. Radar remote sensing is appropriate for monitoring rice, as cultivated areas are most often cloudy and rainy. For this reason, Synthetic Aperture Radar (SAR) is anticipated to be the dominant data source in tropic and sub-tropical regions. SAR satellites can also provide frequent re-visit schedules suitable for agricultural moni-toring. Firstly, this paper presents the results of a study examining the backscatter behavior of rice using multi-temporal RADARSAT dataset.

A rice-type distribu-tion map was produced, showing 4 types of rice with different life spans. The accuracy of rice classification was found to be 91%, providing confidence that multi-temporal RADARSAT data is capable of rice monitoring and has potential for yield estimation. Based on the studies in the Zhaoqing test site, it is suggested that rice yield estimations require three radar data acquisitions taken at 3 stages of crop growth: at the end of the seedling development period, in the ear differentiation period, and at the beginning of the harvest period. Alternatively, if multi-parameter radar data is available, only two data acquisitions are required: at the end of the seedling period, and at the beginning of the harvest period. Secondly, this paper presents the potential of polarimetric SAR technology for rice and tropic vegetation monitoring. A suggestion on optimum SAR system parameter selection for this type of application was made for future system design consideration.

Rice is a heat and water favorite crop. Most paddy rice in the world grows in warm, humid environment with heavy cloud cover and rainfall. It is hard to acquire optical remote sensing data in rice growing regions. Synthetic Aperture Radar (SAR), with all weather, independent of illumination imaging capability and frequent revisit schedule, is anticipated to be the dominant data source for agriculture monitoring in tropic and sub-tropic regions.

Rice monitoring and yield estimation has special significance to China, as rice is the staple grain and accounts for 42% of the crop yield for this country. The estimation of crop yield is a topic of global interest (McDonald and Hall, 1980), and the efficient management of agricultural land resources is strongly related to social and economic sustainable development, especially in China. It is well known that China is the largest country in population. However, as the population increases, and economy and industry develop, the quantity and quality of cultivated land is decreasing rapidly. The food supply to the current 1.2 billion people is a serious concern facing China, and will intensify as the population continues to grow. Therefore, it is important to find an efficient way to face this dilemma. Remote sensing technology will provide the needed information on crop distribution, acreage and potential yield.

This paper presents the results of a study examining the backscatter behavior of rice using multi-temporal RADARSAT dataset. A rice-type distribution map was produced, showing 4 types of rice with different life spans ranging from 80 days, to 120-125 days. Based on previous studies in the Zhaoqing test site, it is suggested that rice yield estimations require three radar data acquisitions. Alternatively, if multi-parameter radar data is available, only two data acquisitions are required.

Test Site and Data Source

The Zhaoqing test site is located in Guangdong Province, south of China center at latitude 22.30, longitude 112.30. It is sited at the northwestern end of Pearl River Delta. The test site was firstly imaged by airborne SAR in 1993 under the GlobeSAR program (Shao, 1995, 1996; Guo 1997). The Shuttle Imaging Radar C-band (SIR-C) and X-band SAR (X-SAR) also flew over the area on April 18, 1994. In addition, there were multiple RADARSAT images acquired from March to December in1996, and from April to July in 1997. The system parameters, imaging modes, and acquisition dates of images used in this study are listed in table 1.

Table 1. System Parameters of SAR Data

Parameters / Data Source	GlobeSAR	SIR-C/X-SAR	RADARSAT (Fine)	RADARSAT (Standard)
Frequency (GHz)	C X 5.30 9.25	L C X 1.24 5.3 9.6	C 5.3	C 5.3
Polarization	HH, HV HH VV, VH VV	HH HH VV HV HV	HH	HH
Incidence Angle (°)	14-45	34.1	43-46(f4)	36-42(S5), 41.46(S6)
Nominal Resolution (m)	6*6	25*25 12.5*12.5	10*10	30*30
Pixel Spacing (m)	6*6	25*25	6.25*6.25	12.5*12.5
Swath Width (km)	18	37.8	50	100
Imaging Date	Nov.20,21,93	April 18,94	Multi-temporal	Multi-temporal

Rice Calendar

In the Zhaoqing test site, there are two crops per year; early season rice and late season rice. There are five major growth periods in the life cycle of rice. 1) Transplanting period: rice plant seedlings are transplanted from the seedbed to the paddy field. The transplanting date depends on the weather, especially temperature; 2) Seedling developing period: the seedling splits up and begins to develop a root system; 3) Ear differentiation period; 4) Heading period: headings begin to form; 5) Mature period: the rice plants mature and are ready to be harvested. Temporally, these five periods for early season rice are March 25-April 5, April 15-25, May 10-30, June 10-25, July 5-31. For late season rice, the growth stages occur as follows: July 20-August 5, August 10-20, September 1-30, October 1-20, November 1-25 respectively.

With progress in agricultural technology, it is known that the longer the rice lifetime, the higher the yield will be. For this reason, the transplanting date of rice is getting earlier. Plant maturity rates and lifetimes are species dependent.

Temporal Backscatter Behavior of Rice

This study aimed at understanding the backscatter behavior of rice over its whole life circle, and the relationship between the rice structure parameters and its backscatter coefficients. Based on a good understanding of backscatter characteristics of rice, we can then suggest the best date for radar data acquisition to monitor the rice growth and estimate the yield. Figures 1 and 2 show the backscatter coefficients of rice extracted from calibrated RADARSAT images acquired from summer to autumn in 1996 and from spring to summer of 1997 respectively. These are produced by applying the calibration procedure to the digital number of Radarsat image to produce s 0 for a training area (20 pixel by 20 pixel) of the multi-temporal RADARSAT dataset.

In figure 4, there are five types of rice: 1) medium mature rice. 2) late mature rice. The backscatter coefficient of late mature rice is slightly higher than medium mature rice on August 4. 3) Late transplanted rice, which has been transplanted about 25 days later than the normal spring rice, due to a cold spring and low temperatures in 1996. The RADARSAT image acquired on April 25 shows the backscatter coefficients of these fields as very low, responding similar to a still water body. The backscatter coefficients from the RADARSAT images taken after April 25 increase gradually, until September when the rice is harvested. Its life span is relatively short, and the backscatter coefficients are lower. 4) single rice and 5) autumn rice.

These two rice types demonstrate how farming activities can change within a year, switch between fish ponds and rice fields. Spring rice was a rice field in spring then switched a fish pond in autumn. Conversely, Autumn rice was a fish pond in spring then switched to rice field in autumn. This practice is very common in Zhaoqing area, as well as many other parts of China. In general, this swap of fish pond and rice implies that the quality of the field is less than excellent, and that the rice yield is normally lower. Based on the classification results of the 1996 multi-temporal fine mode RADARSAT dataset, A land cover map was produced clearly showing the five types of rice, as well as other vegetation covers and targets.

Figure 4. Backscatter intensity of the targets in relation to polarization and frequency (based on the CV-580 data)

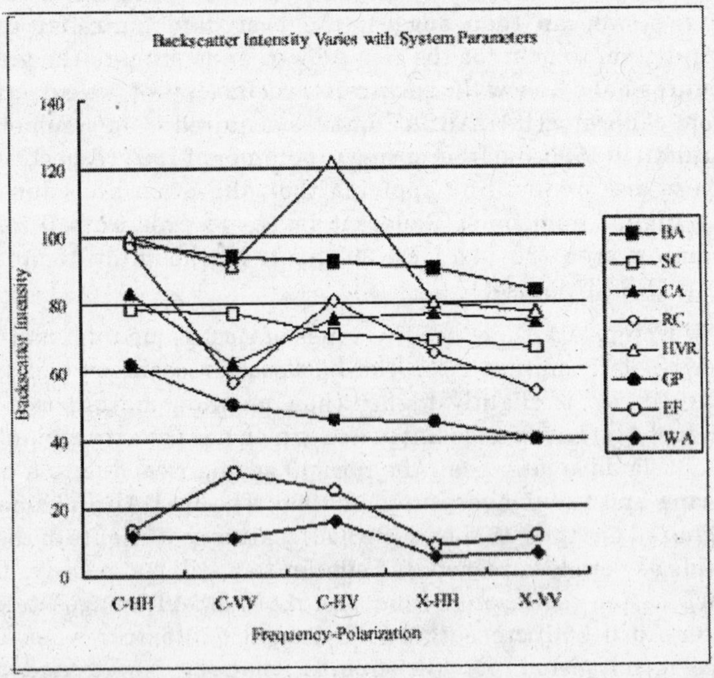

In figure 2, there are four major types of rice with different growth cycles. With the knowledge gained from the rice backscatter behavior of 1996, we can distinguish between rice types quite easily. They are early mature rice, medium mature rice, medium-late mature rice, and late mature rice. Their life span is about 80 days, 100-115 days, 110-120 days and 120-125 days respectively. A false-color fusion image is produced based on the multi-temporal Standard mode RADARSAT images acquired in the spring of 1997. A rice distribution map is produced using the 7 RADARSAT scenes. If a type of rice has higher backscatter value in April, at the early stage or seedling development period, as found with late mature rice, then it was transplanted earlier than the other types of rice. If a type of rice has higher backscatter coefficients in July, the late stage or the harvest period, the rice is harvested later than other type of rice, as found with medium-late mature rice.

From figure 2 we can conclude that late mature rice has the longest lifetime. The medium-late mature rice was transplanted a few days later than medium mature rice and late mature rice, and harvested later. The medium mature rice was harvested before July 22. From figure 2, we can see that early mature rice was harvested before June 28. It has very lower backscatter coefficients since early harvest period. It is concluded that from the early stage, 25 days after transplanting, and late stage, 80 days after transplanting, we can distinguish the rice life span easily, which is an important information for rice yield estimation. The life span is directly related to the yield and quality of rice. This information is provided by local agronomists. It has been proved in real life. The farmer used to grow three rice crops a year in past, nowadays, they only grow two rice crops a year and they make more money from rice farming.

Based on figure 2, the knowledge we had in our research started in 1993, we established the theoretic backscatter model of rice. Figure 3 mainly represents the backscatter behavior of medium-late mature rice produced by the theoretic backscatter model. It is transplanted in April 1 and harvested in July 30. The late mature rice and medium mature rice are transplanted earlier than medium-late mature rice. Medium mature rice is harvested earlier than medium-late mature rice. This model can be applied to other type of rice with the information on rice life span, transplant date and harvest date. Then we can calculate the backscatter coefficients of rice at certain growth stage.

The results from the studies carried out in 1996 and 1997 imply that the most important radar data acquisition time for rice monitoring is at the end of the seedling period and ear differentiation period, which is late April for spring rice and late August for autumn rice. For calculating the acreage of planted rice, two radar data acquisitions are required: one near the end of April, and another in the middle of May during the ear differentiation period. For rice yield estimation, three radar data acquisitions are required: one at the end of April, one in the middle of May, and one at the end of June. For autumn rice or late season rice, the corresponding date can be found by checking the rice growth calendar.

A rice-type distribution map was produced, showing 4 types of rice with different life spans ranging from 80 days, to 120-125 days. The life span of a rice crop has significant impact on yield, as well as the taste and quality of the rice. The yield of three counties and two administrative regions, totaling 5000 square kilometers, are estimated in this study. The accuracy was found to be 91%, providing confidence that multi-temporal RADARSAT data is capable of rice monitoring and yield estimation. Based on previous studies in the Zhaoqing test site, it is suggested that rice yield estimations require three radar data acquisitions taken at 3 stages of crop growth: at the end of the seedling development period, in the ear differentiation period, and at the beginning of the harvest period. Alternatively, if multi-parameter radar data is available, only two data acquisitions are required: at the end of the seedling period, and at the beginning of the harvest period.

Optimum SAR System Parameter for Rice and Tropic Vegetation Monitoring

Figure 4 and 5 were delivered from Canadian airborne radar system-CV-580, with X and C band, quad-polarization and USA SIR-C/X-SAR with X, C and L band, HH, HV, VV polarization. These two figures show great advantage of multi-frequency, multi-polarization radar system in rice and tropic vegetation discrimination. The detail will be made in presentation.

Figure 5. Comparison of intensity of rice and water in relation to polarization and frequency (based on SIR-C/X-SAR data)

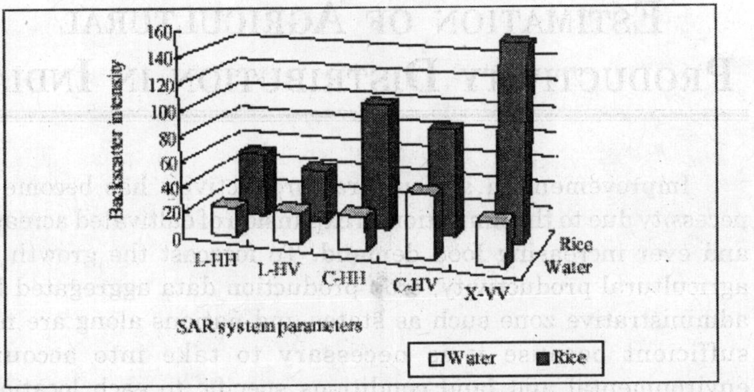

It is suggested that the optimum SAR system parameter for rice and tropic vegetation monitoring should be C and L band, polarimetric is favorite if it is possible, if not, HH, VV or HV polarization are the least, 10-25 meter spatial resolution, 8-16 days revisit schedule, 35-45 incidence angle, as wide swath as possible to cover large area and bring the cost down.

3

Estimation of Agricultural Productivity Distribution in India

Improvement in agricultural productivity has become a necessity due to the limitation in expansion of cultivated acreage and ever increasing food demand. To forecast the growth of agricultural productivity, crop production data aggregated by administrative zone such as states and nations along are not sufficient because it is necessary to take into account environmental and land conditions specific to each location. With this background, this paper discusses a method to estimate the distribution of agricultural productivity as case study of India. The distribution of NPP for each grid cell (8km resolution) was estimated using the NDVI data derived from NOAA AVHRR and PAR. NPP in crop land was extracted using the land use data and has been summed up by each districts. Correlation of the agricultural productivity and NPP were investigated by comparing statistical crop production data on each district. As the result, it was found estimated from remote sensing data is effective to complement data of crop production. Finally the grid-level agricultural productivity was estimated using NPP data.

INTRODUCTION

Food demand is foreseen to continue increasing along with the increase in the world population. However, an expansion of cultivated acreage can not be expected because of the limitations arising from environment conservation, especially, forest conservation. Therefore, an improvement in farm productivity has become a necessary for coming days. The growth of the agricultural productivity depends on the land conditions, for example availability of irrigation water resources, soil deterioration, in addition to an improvement of agricultural technology. Mechanization is also influenced by the topography. Whereas, increase in productivity by applying fertilizer are limited due to their side effect of environmental loading. Like this, to foreseen the growth of the agricultural productivity, various environmental factors must be considered, and grasping on detailed space distribution of the crop production becomes indispensable. Moreover, monitoring of the farming area in cultivating stage is effective, because the amount of harvest is insufficient. So remote sensing technology can be very effective for spatio-temporal supplementation.

The purpose of this study is to develop a basic method to estimate distribution of the agricultural productivity using remote sensing data, gives an approach to a fundamental research. India has been selected as study area, because the ratio of the crop land is relatively higher and availability of statistical data.

To do this, NPP distribution (8km resolution grid cell) was estimated using NDVI of NOAA AVHRR and PAR data derived from TOMS data set. Next, NPP on crop land (named "crop land NPP") was extracted using the land use data, and totaled in each district. The distribution of the agricultural productivity was estimated, by corresponding with crop land NPP and the statistical data of crop production.

ESTIMATION OF NPP DISTRIBUTION USING NDVI

NPP distribution was estimated with a method proposed by Hirakoba (1998) based on Production Efficiency Approach. NPP is estimated from solar radiation, absorbed by a plant, and the efficiency index (e) which is the ratio of radiation energy changed into the organic substances.

The radiation energy in wavelength which a plant can use for the photosynthesis is called PAR (Photosynthetically Active Radiation) and data set of PAR was made using the data on TOMS (Dye, 1993). The ratio of PAR absorbed by the canopy of the plant is named fapar (the fraction of PAR), and has been estimated using NDVI of NOAA AVHRR. Solar radiation absorbed by a plant is named APAR (Absorbed PAR) and have been calculated by multiplying PAR and fapar. Therefore, APAR can be calculated from the data set of PAR and NDVI.

In this study, using 8km resolution monthly NDVI data derived from NOAA AVHRR data (figure 1), APAR was estimated. Moreover, monthly NPP was estimated using APAR multiplied by efficiency index (e) which was assumed 1.5 g/MJ, and annual accumulation value was calculated. The NPP distribution estimated in this study is shown in figure 2. Crop land NPP was extracted from NPP distribution using the land use data (figure 3).

Figure 1: Monthly NDVI (example of March, June, September, December)

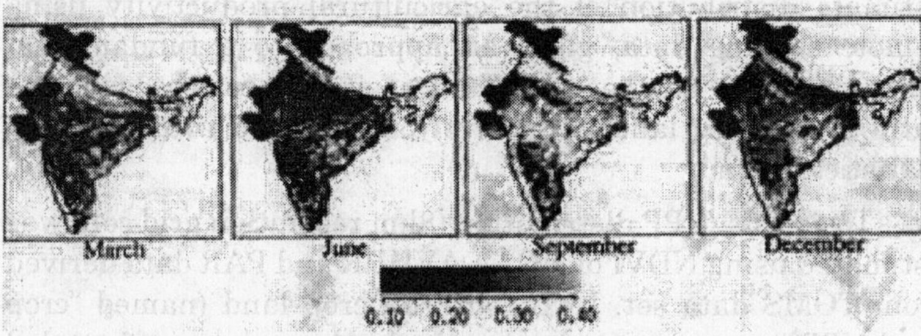

Figure 2: NPP distribution (annual, 1990)

Figure 3: Crop land NPP distribution

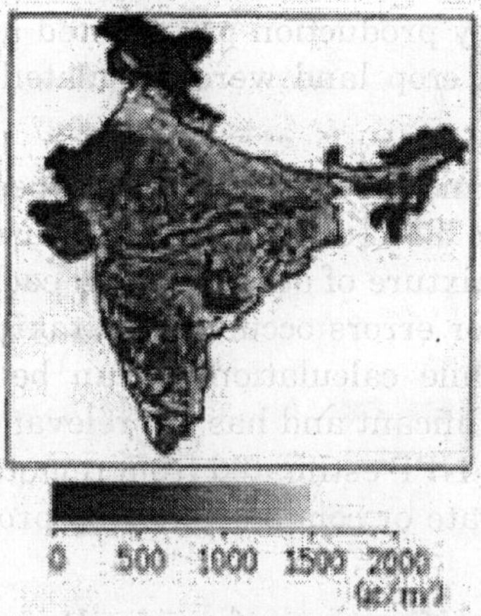

NDVI data for the year 1990 was used to match with the statistical data. PAR data available with us was only for the year 1979-1989, therefore monthly mean value of PAR data in this period was calculated and used.

Investigating Correlation between Agricultural Productivity and NPP

The total amount of the organic substance, produced in the crop land, was estimated by multiplying crop land NPP (as describe in proceeding section) and the area under each grid cell. Moreover, verification was tried by calculating them in each district boundary and their correspondence was made with statistical data of crop production. But, in reality is not possible because, crop land NPP is total amount of all crop types whereas statistical data available are on crop basis. Also, the rates of NPP produced in crop land changed into the crop were different by their types. Therefore, it is inappropriate to calculate the total amount of crop production of all crop types. Hence, districts were classified into respective agriculture areas, they were examined for each agricultural land. Rice, maize, wheat, jowar, bajra were selected as object and as they are the major crops in India. The classification into agricultural area was done by selecting the crop planted as shown in the figure 5 and correlation coefficient of crop land NPP and the amount of crop production is shown in the table 1. But, to get rid of the correlation which occur from the difference in the area of each district, they production per planted area and crop land NPP per area of crop land were calculated and used for the purpose.

The significance of difference crop types can be seen from table 1 and figure 5. Actually, many crops are produced in each district, so the mixture of other crops in each agriculture area are the reason for errors occurred. So taking the factors into consideration while calculation, it can be said that results obtained are significant and has got relevancy. Therefore, this results says that NPP estimated from remote sensing data can be used to estimate or complement crop production to a large extent.

Figure 4. Classification of agriculture area based on planted area

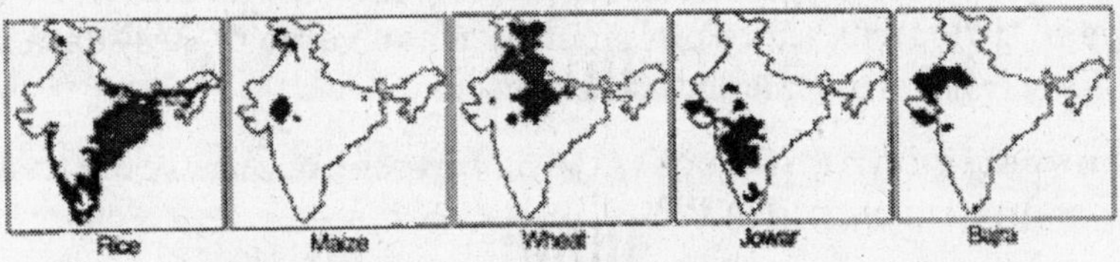

Figure 5: The scatter diagram of crop land NPP and the amount of crop productivity

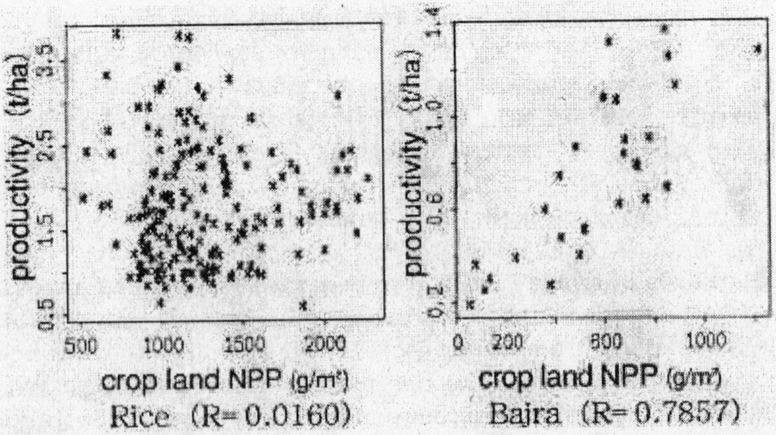

Table 1. correlation coefficient of crop land NPP and the amount of crop production

Rice	R=0.0160
Maize	R=0.4801*
Wheat	R=0.2748**
Jowar	R=0.5451**
Bajra	R=0.7857

**...Significance level 1%*...Significance level 5%

Estimation of Agricultural Productivity Distribution

The distribution of the agricultural productivity was estimated using above examination result. This is shown in figure 6. These were calculated using the agricultural land NPP to complement data available on crop production of each district.

Figure 6. Agricultural productivity distribution

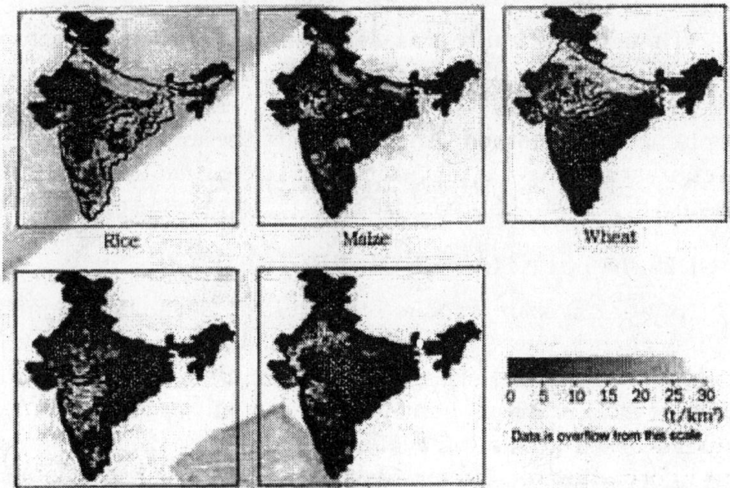

These spatial distribution of the agricultural productivity are open to evaluate the limitation of the environmental conditions and can also be used to examine the potential production of crop land by overlaying other geographical data such as climate and hydrology. Moreover, this research gave the possibility that is can be applied to the prediction of the damage in the disaster mitigation and so on.

CONCLUSIONS

In this study, the distribution of the agriculture productivity in India was estimated using remote sensing. The conclusions are as follows.

1. NPP distribution and crop and NPP distribution in India were estimated by Production Efficiency Approach using remote sensing data.
2. The correlation of the agricultural productivity and NPP is investigated and it was found that NPP estimated from remote sensing data is effective to complement data of crop production.
3. The distribution of the agricultural productivity was estimated using NPP derived from remote sensing data and the statistical data of crop production.

Though this study was kept within case study of India, there is possibility that agricultural productivity of the area where statistical data aren't prepared can be estimated by applying same method. And to predict the amount of harvest in cultivating stage, application of real time remote those future studies, the result obtained from this study seems quite useful.

MODELING SPATIAL CROP PRODUCTION

Spatial Erosion Productivity Impact Calculator (Spatial-EPIC) gave a new direction to simulate crop production at regional scale based on microscopic simulation at each small piece of land in an efficient way. The result presented in this paper is based on existing/available spatial agro-climatic dataset but information related to soil and water has not been included spatially at this stage as they are under process of development. Using Spatial-EPIC crop model observed crop yield for the period 1995-1993 are compared with yields simulated and it is found that agreement between simulated and observed crop yield shows better correlation. GIS based modeling of an agroecosystem is expected to give a new approach in order to provide agricultural managers with a powerful tool to assess simultaneously the effect of farm practices to crop practices to crop production in addition to soil and water resources. At present, most of the crop models are location explicit information. Therefore, development of spatially or raster based biophysical crop model with go a long way in helping us to understand many intricacies of modeling of large areas at 1 to 10 km resolution. The large-scale distribution of crops is larges determined by climate. This study gives a preliminary result of Spatial-EPIC (which is in the process in identifying weakness of GIS data interpolation capabilities. We focused our efforts mostly on climatic dataset generation for three major crops Maize-Wheat-Rice which can be further used to analyze the response of other natural systems to climate change in a later course. The scale resolution of most crop models used in simulating climate change impacts is, for all practical purposes, a single point on the earth's surface. To get regional estimates of crop responses to climate change requires either scaling up the model simulations from point estimates or scaling up the data inputs from point measurements and then performing the simulations. Both ways will result in some loss of information due to aggregation error. The effect of spatial scale of climate inputs on crop simulation is examined in this study.

The purpose of this paper is to examine the effect of spatial aggregation (i.e, the size of the areas from which the synthesis of multiple in situ data points into a single areal estimate is made) of climate data used in crop models and find an agreement between modeled and observed crop yields. We begin with interpolation of WMO data at the degree (almost 10km on ground) grid resolution. Specifically this paper evaluates the relative impacts of climate and crop spatially throughout the study area.

Materials and Methods

In this section, the followings are described; the region of the study, 1) the crop model used in the analysis, 2) the modeling of climate data to a climate model grid network and 3) the procedures used to compare observed and simulated yields over a continuum of spatial scales within the grid network. The area chosen for this study compasses one of the northeastern state of India namely Bihar. The area studied is located between 21° 58' to 27° 31' North latitude and 83° 19' to 88° 17' East longitude. It is overlain at 10km grid size, which cover the entire state Bihar (Figure 1). Three different crops were modeled for the period of 1985-1993 using (Spatial-EPIC) a crop growth model described below. In northern India there are two distinct seasons Kharif (July to October), and Rabi (October to March). Therefore the crops modeled were Kharif maize, Rabi wheat and kharif rice at every two year crop rotation. The period between March to June (locally called as Zaid season) left fallow during the simulation. This distribution mimics the predominant crops, which is being taken widely in the area.

Figure 1. Study area showing WMO stations and overlain with interpolated climate data as an

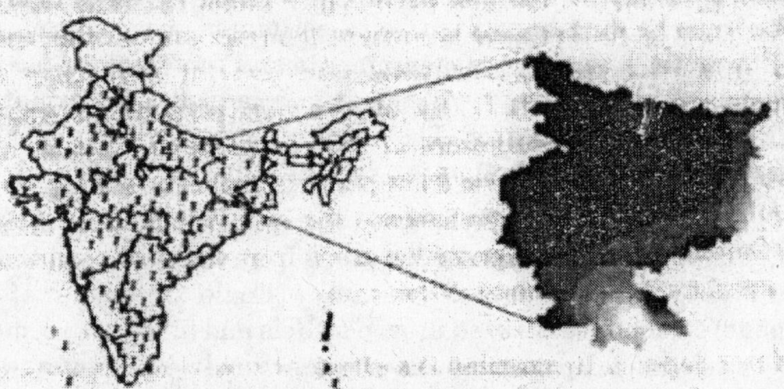

Figure 1. Study area showing WMO stations and overlain with interpolated climate data as an

Simulation Models

Until now there have been a lot of studies on agricultural potential productivity but to relate actual crop productivity, model based simulations are not adequate. Estimates of on-farm and off-farm productivity are being done using experimental/point based model. Biophysical spatial based model is still lacking to compute them at regional or national level. Simulation models, such as EPIC (Willimas, 1990) model used was originally point-based developed by USDA-ARS. The first results of Spatial-EPIC being developed by the author is presented in paper (described below) together with mathematical presentation of the biophysical crop simulation process. The model is designed to predict what would happen under different conditions, such as change in temperature, precipitation, atmospheric CO2, soil moisture and so on at regional/national scale. The results presented in this paper are based on climate data. A model developed by Richardson (1981) was selected for use,in EPIC because it simulates temperature and radiation, which are mutually correlated with rainfall. The weather generator module offers a convenient method of obtaining the numerous long sequence weather data that are required for the simulate

Spatial-EPIC

Spatial-SPIC is a crop simulation model being developed to estimate the relationship between soil erosion and crop productivity, which has been implemented in GIS environment to have spatial distribution of crop output. In addition to the plant growth routine, the model includes components for weather simulation, hydrology and crop management. It operates on a daily time-step. Among factors simulated by EPIC are evapotranspiration (based on Penmen Monetih model), soil temperature, crop potential growth, constraints and yields. EPIC uses a ssingle model for simulating all crops, although, of course, each crop has unique values for model parameters. For this study, CO2 concentration held at current ambient level (350ppm) has been used. Crop yields are estimated by multiplying the aboveground biomass at maturity (determined by accumulation of heat units of specific field harvest data) by a harvest index for the particular crop.

Preparation of spatial-EPIC Dataset

Model requires data on soil properties, which has been assumed same for all the area as Spatial-EPIC is under progress and data required for these parameters are under development. The detailed profiles of relevant production characteristics (for example, fertilization, planting, harvesting, irrigation and heat units for maturity) were complied for the crops simulated across the region. The range of values for those production characteristics is shown on table 1. the weather variables necessary for those production characteristics is shown on table 1.

The weather variables necessary for driving the EPIC, a main impetus of this paper, are daily values of precipitation, minimum/maximum are temperature, solar radiation and humidity. A detailed discussion of climatic data generation is described in the next section.

Table 1. Management practices used in the model

Maize	Planted Total Fertilizer(urea+elem.-Phosphorus)	7th July 40+10 kg
	Total no. of irrigation/amount, (500/800)	2/1300 mm
Wheat	Sown	10th November
	Total ertilizer(urea+elem.-Phosphorus) Total no. of irrigation/amount, in 3 splits	75+15 kg 3/3000 mm
RiceWithout Irrigation	Planted Total Fertilizer (urea+elem.-Phosphorus)	7th July 40 +10 mg

Data Generation: Interpolation and Interpretatio

Most simulation model assumes homogeneous conditions over the space they are representing. In general, these conditions do not exceed a few square kilometers in extent. To input weather data across India, for example, we were faced with the problem of extending required model input over several hundred square kilometers between weather stations.

Several interpolation procedures are available, ranging from simple linear interpolation techniques and triangulated networks, to more sophisticated distance weighing or kriging techniques. It must be remembered that the only information used by any interpolation procedure is the location of known values. For this study, the climatic information used to make interpolation is based on inverse distance weighted method of more than 200 stations in India from which Bihar has been delineated for the purpose. While interpolation of a value at reach cell in the study area using 200 meteorological stations is technically easy, but some important questions can be raised : are the stations representative of the areas around them? How large or small an area do they represent? How does the spatial feature in question change over space? Is it continuos or discontinuous and abrupt?

We used same climate dataset for the different maps resolution across study area. Depending on its cell size, the level of detail becomes available at different resolution. It is thus important to realize the size of the dataset used to create GIS maps (as model input) and is important for model application from its accuracy and better performance view point. On the other hand a total of one or two stations falling in each administrative boundary may not be enough to give great confidence in the resulting interpolated map.

Results

Climatic data at 10 km spatial aggregation needs to be evaluated to identify its applicability in crop modeling; result of those are presented here. Observed yields were compared against simulated yields for two different grid cells of these there crops in the study area. As a mapping tools, GIS has been used for the production of descriptive maps using simulation model outputs that portray the spatial distribution of model prediction and the spatial juxtaposition of specified subsets of the data.

The geographical locations of the modeled process of different crops are simulated actual yield but there is a lack on observed data to tie the results to a real world location. Figure 2 shows an example of comparing spatially distributed simulated crop yield with observed aggregated crop yield at district levels. The above results suggest that the aggregation of climate data alone showed pertinent agreement between observed and simulated yield, as the other parameters except climate remain unchanged. Observed data available for validation are average of district administrative boundary, whereas simulated output varies cell by cell as of their resolution. To check its validity more explicitly we averaged the simulated yield at district level as zonal mean productivity and then compared with observed average. Figure 3 shows a sample of linear regression between observed vs simulated crop yield of rice, wheat and maize in the year 86, 89 and 91 respectively.

The outputs obtained on climate data gave quite satisfactory result and its is supposed that after introduction soil, topography and other input parameters spatially the model will yield a better result.

Figure 2. Comparing Spatially Distributed Simulated Crop Yield vs Aggregated Observed Crop Yield

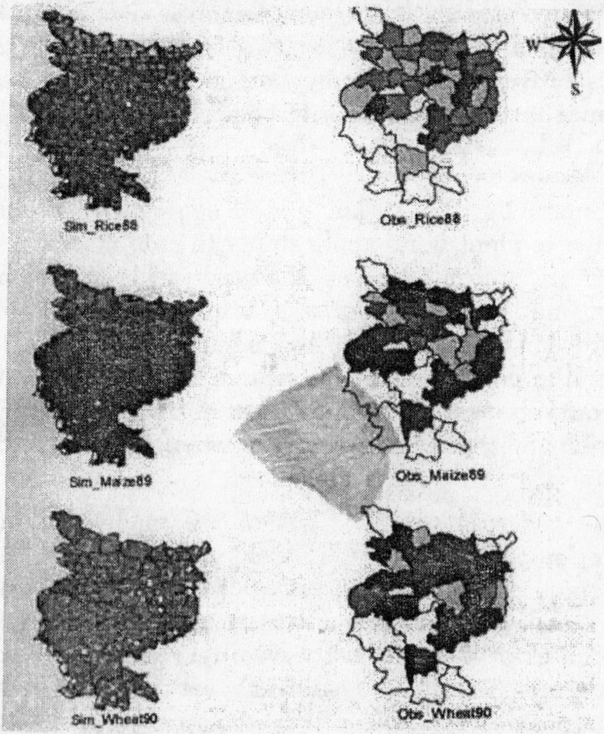

Figure 3. Linear regression of observed vs. simulated crop yields

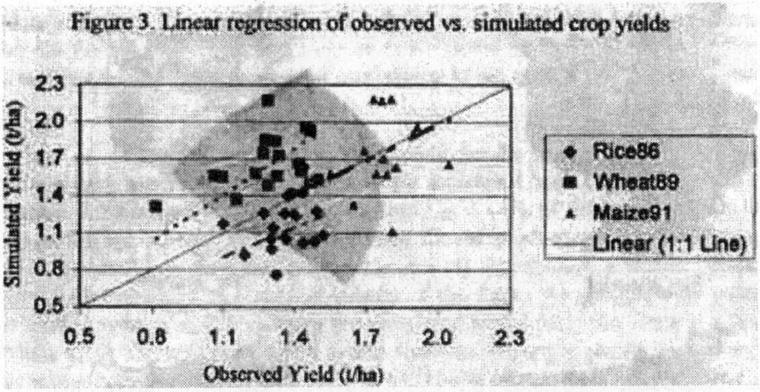

Therefore, result reinforces the conclusion that in the range of spatial output examined here, agreement between observed maize, wheat and reice with simulated Spatial-EPIC model is sensitive and it depends a lot on the climate data.

Analytic operation allow GIS to generate new data correlation and prediction of future conditions. Research is still being conducted and preliminary results from this study are just becoming available.

However, it is encouraging to have found a methodology that Provides a plant physiologist, a modeler and a GIS users a common ground to discuss simulation results and further potential research directions.

Simulated crop yield maps generated can be used to better communicate model predictions. Hence using this methodology a region/nation can be modeled for crop productivity, which help researchers and decision-makers understand the status and extent of climate effects on global processes such as rice, wheat and maize production.

As of now, the most difficult problem countered in applying Spatial-EPIC areas in the developing world is the apparent lack of quality environmental data availability, detail validity, comprehensiveness, and scale must be carefully considered before this methodology can be successfully implemented.

CORRELATION OF SPECTRAL REFLECTANCE TO GROWTH IN RICE VEGETATION

Multispectral data from aircraft or satellite have been widely used in agriculture as well as in other industrial sectors. To connect such data with vegetation cover for agricultural applications, generally a detailed and through ground-based should be made to characterize radiance and reflectance spectra and link it with growth and yield traits. This study was to remotely measure and monitor the seasonal variations in reflectance spectrum, in the range of 350-1100 nm, of rice (Oryza sativa L. cv. Tainung 67) canopy at the experimental field of Taiwan Agricultural Research Institute (TARI) during the cropping seasons in 1996-1997. The objectives were to examine relationship between growth traits and NDVI and to assess their potential use in estimating growth of rice crop. Attempts were also made to compare NDVI calculated from narrow-band ground-based spectral measurement and the simulated board-band multispectral satellite data. The physiological and morphological implications of the relationships were discussed.

Satellite or aircraft remote sensing can provide a timely spatial distribution information on crop performance and be incorporated into farm management scheme. Application such as precision farming activities, growth and production estimation, and land surface characteristics evaluation are made possible. Before these practical applications, vegetation spectral properties from remote sensing must be in connection with some physical characters of plant canopy so that plant performance can reasonably be assessed. Mathematical formulae are then established from the fitting curves of these paired variables. Leaf area index (LAI), percent vegetation cover, and unit area plant biomass are among physical characters commonly used as indicators of the status of crops.

Vegetation indices are mathematical transformations intended to estimate the spectral contribution of crop vegetation to multispectral observations. The transformations have the effect of normalizing measurements acquired from varied environmental conditions. The formulae are derived mostly from discrete green, red and near-infrared bands, especially the two ends of the so-called chlorophyll red-edge. By correlating vegetation indices with physical characters of plant canopy, changes of vegetation feature can be potentially assessed and predicted from values of vegetation indices during the growing season.

In this study, variations in reflectance spectrum were monitored and analyzed and the normalized difference vegetation index (NDVI) were calculated from the ground-based remotely sensed spectral measurements of rice canopy for the growing seasons of 1996-1997. An simplified vegetation model for estimating crop growth traits from NDVI is presented, and the simulated broad-band versions of NDVI were compared with the narrow-band counterparts.

Materials and Methods

Experiments were conducted at TARI Experimental Farm (23o30'N, 120o42' E, elevation of 85 m) on a loam soil (nonacid, hyperthermic, Fluvaquentic Dystrochrept) during First and Second crops of 1996 and 1997. Cultivation and fertilization were in consistent with the existing cultural practices for rice adopted by the local region. Each season had 3 plots as replications and were furrow-irrigated. Herbicide and hand-weeding were practiced to avoid weed interference. Plants were sampled from 2 wk after transplanting till harvest periodically. Leaf area was determined and LAI, defined as the area of green leaves per unit area of land, was computed at each sampling. Leaf dry weight (LDW) was measured after oven-dried at 80oC for 72 h. Means were used for regression analyses to examine the correlation between growth traits and NDVI.

Method of spectral measurements was reported in the previous papers (Yang and Su, 1997, 1998). Spectral irradiance of the dense vegetation cover and incident solar radiation in the range of 350 to 1100 nm was measured using a LI-1800 high spectral resolution portable spectroadiometer (LI-COR inc., Nebraska, USA). The sensor head was pointed downward in a nadir-viewing and was placed horizontally 1.0 m above rice vegetation surface to scan the upward reflected radiation.

Measurements were made at each plot using sunlight as illumination source on clear or near cloudless days between 11:00 to 12:00 local standard time. Date were averaged and mean values were used.

Radiance spectra of sunlight were taken before and after each replicate measurement and the average was used as the reflectance standard. Reflectance of individual wavelength was calculated by dividing the vegetation radiance measurements with the corresponding incident solar radiation measurement with the corresponding incident solar radiation measurements. Reflectance spectra measured during the growth were grouped for each crop and the standard errors of individual wavelengths were calculated as representative of the upper and lower boundaries. The NDVI was defined as (NIR-RED)/(NIR+RED). The reflectance of the wavelengths at RED (674 nm) and NIR (754 nm) measured from LI-1800 were selected and used for calculations. Regression analysis were performed to generate fitting-curves in order to monitoring seasonal changes of NDVI as well as growth traits.

From narrow-band reflectance spectra, broad-band RED and NIR reflectance were computed. The reflectance simulated for RED and NIR for HRV sensors of French SPOT program were derived from the 2-nm-wide narrow-band LI-1800 spectra in bandpasses of 610-680 nm and 790-890 nm. For MSS sensor system of Landsat series, bandpasses in the range of 600-700 nm and 700-800 nm were computed to simulated Band 2 and Band 3, respectively. The averaged values were presented for RED and NIR bands of the broad-band sensors. Linear regression was performed to compare the difference in NDVI between values calculated from single reflectance and band reflectance.

Results and Discussion

For a typical reflectance spectrum, wavebands in the visible region were mostly absorbed by chlorphylls while wavelengths in NIR were reflected significantly because of the absence of absorption (Gausman st al., 1969). It was shown similar in seasoanl pattern of reflectance spectrum but was shifted by weather and soil impact (Fig 1). Variations were found much greater in the early growth and maturing stages than during the long middle period of the growing season. The greater standard errors graphed in the 1st crop of 1997 inferring a stronger climate oscillation occuring in the growing period. These 'environmental' and 'growth' effects influencing the time-seqential performance of canopy reflectance spectra were also found in many field crops (Masoni st al., 1996; Sinclair et al., 1971; Yang Ko, 1998).

Figure 1: Seasonal variations in reflectance spectra of rice vegetation cover over the cropping seasons of 1996 and 1997. The upper and lower levels are the intervals of standard errors.

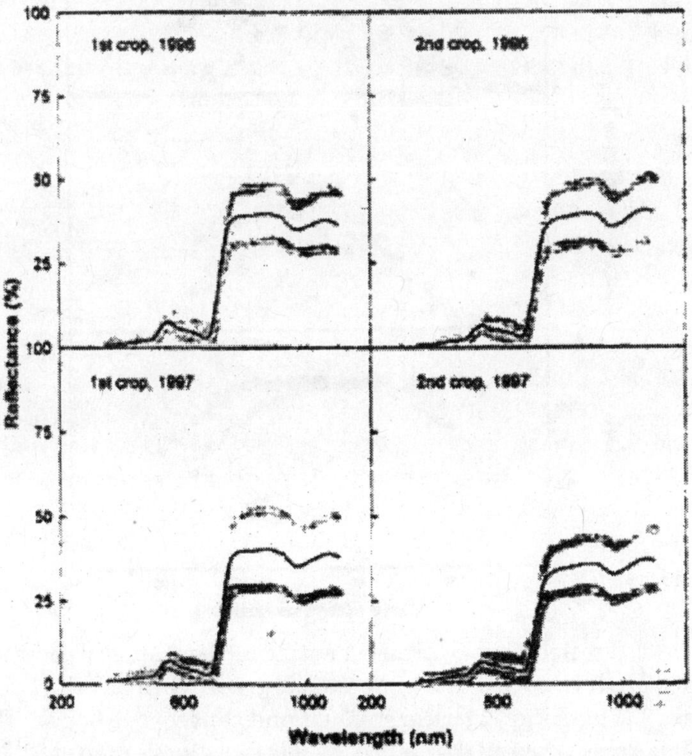

Fig. 2 shows that NDVI curve reached the peak near heading, and then fell off as the growing season continued. It can be accounted for by the seasonal trends of reflectance at 674 and 764 nm, the increasing or decreasing difference between RED and NIR reflectance giving a corresponding upward or downstream NDVI.

Figure 2: Seasonal changes in leaf area index (LAI) and leaf dry weight and the normalized difference vegetation index (NDVI) of rice vegetation over the cropping seasons of 1996 and 1997.

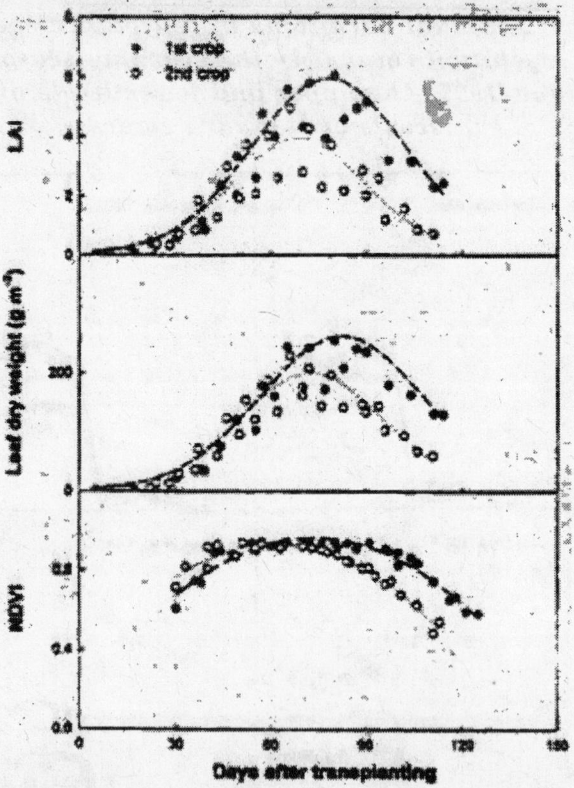

RED reflectance decreased with time beacause of increased chlorophyll absorption by increased green vegatation (Tucker et al., 1979). As growth diminished and senescence began, RED reflectance started to increase. It was reversed for NIR reflectance. The NIR bands are essentially unabsorbed by foliage and are mostly transmitted and reflected with vegetation structure (Sinclair st al., 1971). Ashley and Rea (1975) observed the similar results following the ground foliage production and used it to depict phennological change.

Figure 3: The curvilinear relationships between leaf area index (LAI) and leaf dry weight and the normalized difference vegetation index (NDVI) of rice vegetation over the cropping seasons of 1996 and 1997.

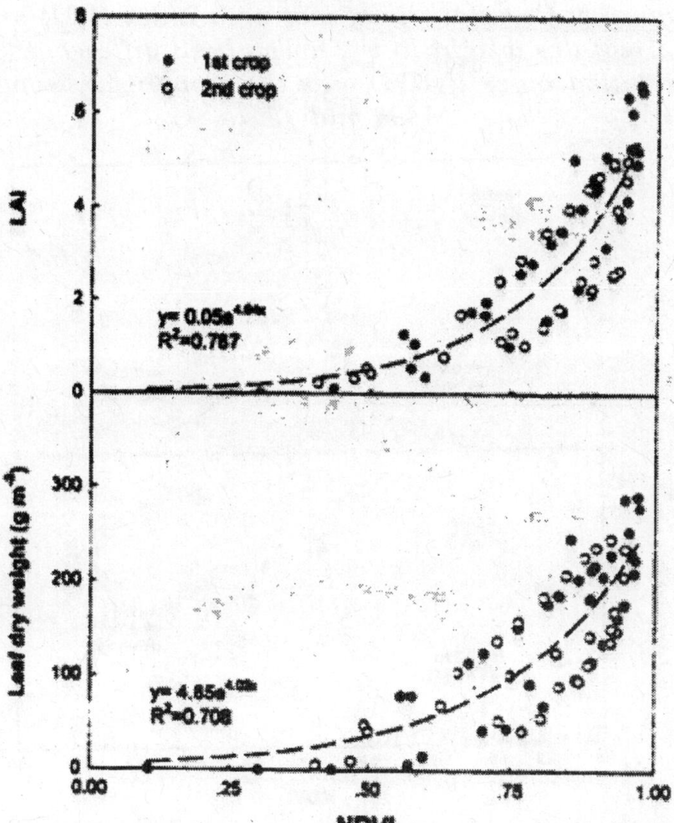

The increase and decrease of NDVI of NDVI followed the changes of growth traits (Fig. 2). It suggests that NDVI is sensitive to the amount of photosynthetically active vegetation and may be a superior variable for monitoring vegetation. Fig. 3 further shows that relations of these growth traits and

NDVI were height. Changes of NDVI agree well with the cnages of canopy cover (up to LAI of ca. 5.5) and foliage biomass (up to LDW of ca. 240 gm-2), indication a strong association of percent vegetation cover and plant chlorosis with NDVI. Tucker (1979) had a similar result, showing that NDVI was sensitive primarily to the green leaf area of green leaf biomass. Curran (1983) estimated LAI from RED and NIR reflectance measurements by correlating LAI with ratio vegetation index.

Figure 4: Correlations of leaf area index (LAI) and leaf dry weight to the normalized difference vegetation index (NDVI) over the cropping seasons of 1996 and 1997.

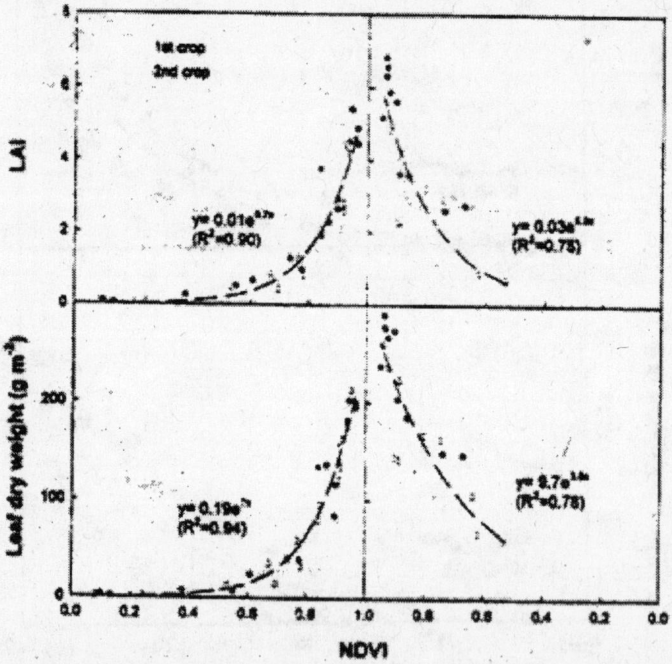

By separating NDVI into an uprising growth part (vegetative phase) and a downsizing maturing part (reproductive phase), growth transition are distincted from the maximum point of NDVI (Fig. 4). While for a given value of growth trait, there are two corresponding points of NDVI be measured at two different growth phase. In the vegetative phase, the rapid increase in vegetative canopy prior to heading was manifested by an exponential increase in NDVI.

Figure 5: The linear relationships between NDVI calculated from band reflectance (NDVI^{SPOT} and NDVI^{NDVI} and NDVI calculated from single reflectance NDVI ^{Single}

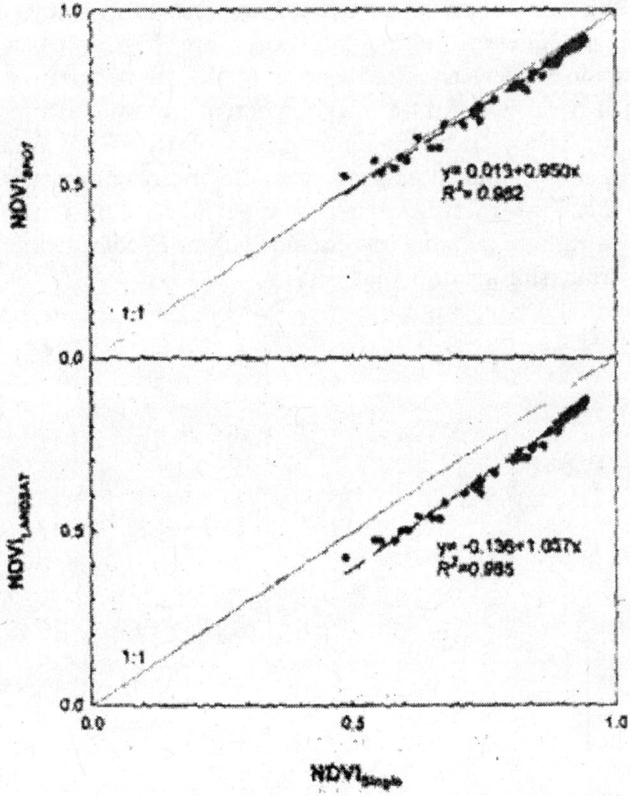

The NDVI reached a plateau when canopy cover was complete. During late season, decline of NDVI was responsive to the onset of chlorosis due to crop maturation. For a précised estimation of canopy status, relationship between growth traits and NDVI should be carefully classified. Because the relationship are adversely influenced by senescence, it would be better not to assess growth traits in the early and the late stages of plant development (Ahlrichs and Bauer, 1983). As the sustainability of full vegetative cover is an important factor in determining the final yield, the spectral assessment of the development stages is crucial (Tucker et al., 1979).

To examine whether satellite broad-band data can be used to substitute for ground-based narrow-band measurements, a linear regression analysis was performed (Fig. 5). The broad-band versions of NDVI agree very well (R^2=0.982 for SPOT and R^2=0.985 for LANDSAT) with the narrow-band counterparts. As these spectral transformations are highly correlated, estimation of growth traits based on NDVI can be fairly accurate for both narrow-band data and spectral inputs collected from satellite. Thus, the reflectance band of the HRV and MSS sensors could be used as predictors of canopy characters. With many more interacting variables, spectral data measured from satellite remote sensing are considered more complicated than those from the ground platform.

4

Industrial Agriculture

Industrial agriculture is a form of modern farming that refers to the industrialized production of livestock, poultry, fish, and crops. The methods of industrial agriculture are technoscientific, economic, and political. They include innovation in agricultural machinery and farming methods, genetic technology, techniques for achieving economies of scale in production, the creation of new markets for consumption, the application of patent protection to genetic information, and global trade. These methods are widespread in developed nations and increasingly prevalent worldwide. Most of the meat, dairy, eggs, fruits, and vegetables available in supermarkets are produced using these methods of industrial agriculture.

HISTORICAL DEVELOPMENT AND FUTURE PROSPECTS

The birth of industrial agriculture more or less coincides with that of the Industrial Revolution in general. The identification of nitrogen, potassium, and phosphorus (referred to by the acronym NPK) as critical factors in plant growth led to the manufacture of synthetic fertilizers, making possible more intensive types of agriculture. The discovery of vitamins and their role in animal nutrition, in the first two decades of the 20th century, led to vitamin supplements, which in the 1920s allowed certain livestock to be raised indoors, reducing their exposure to adverse natural elements. The discovery of antibiotics and vaccines facilitated raising livestock in concentrated, controlled animal feed operations by reducing diseases caused by crowding. Chemicals developed for use in

World War II gave rise to synthetic pesticides. Developments in shipping networks and technology have made long-distance distribution of agricultural produce feasible.

Agricultural production across the world doubled four times between 1820 and 1975 to feed a global population of one billion human beings in 1800 and 6.5 billion in 2002. During the same period, the number of people involved in farming dropped as the process became more automated. In the 1930s, 24 percent of the American population worked in agriculture compared to 1.5 percent in 2002; in 1940, each farm worker supplied 11 consumers, whereas in 2002, each worker supplied 90 consumers. The number of farms has also decreased, and their ownership is more concentrated. In the U.S., four companies kill 81 percent of cows, 73 percent of sheep, 57 percent of pigs, and produce 50 percent of chickens, cited as an example of "vertical integration" by the president of the U.S. National Farmers' Union. In 1967, there were one million pig farms in America; as of 2002, there were 114,000, with 80 million pigs (out of 95 million) killed each year on factory farms, according to the U.S. National Pork Producers Council. According to the Worldwatch Institute, 74 percent of the world's poultry, 43 percent of beef, and 68 percent of eggs are produced this way.

According to Denis Avery of the agribusiness funded Hudson Institute, Asia increased its consumption of pork by 18 million tons in the 1990s. As of 1997, the world had a stock of 900 million pigs, which Avery predicts will rise to 2.5 billion pigs by 2050. He told the College of Natural Resources at the University of California, Berkeley that three billion pigs will thereafter be needed annually to meet demand. He writes: "For the sake of the environment, we had better hope those hogs are raised in big, efficient confinement systems."

BRITISH AGRICULTURAL REVOLUTION

The British agricultural revolution describes a period of agricultural development in Britain between the 16th century and the mid-19th century, which saw a massive increase in agricultural productivity and net output. This in turn supported unprecedented population growth, freeing up a significant percentage of the workforce, and thereby helped drive the Industrial Revolution. How this came about is not entirely clear. In recent decades, historians cited four key changes in agricultural practices, enclosure, mechanization, four-field crop rotation, and selective breeding, and gave credit to a relatively few individuals.

Challenges and Issues

The challenges and issues of industrial agriculture for global and local society, for the industrial agriculture industry, for the individual industrial agriculture farm, and for animal rights include the costs and benefits of both current practices and proposed changes to those practices. Current industrial agriculture practices are temporarily increasing the carrying capacity of the Earth for humans while slowly destroying the *long term* carrying capacity of the earth for humans necessitating a shift to a sustainable agriculture form of industrial agriculture. This is a continuation of thousands of years of the invention and use of technologies in feeding ever growing populations.

[W]hen hunter-gatherers with growing populations depleted the stocks of game and wild foods across the Near East, they were forced to introduce agriculture. But agriculture brought much longer hours of work and a less rich diet than hunter-gatherers enjoyed. Further population growth among shifting slash-and-burn farmers led to shorter fallow periods, falling yields and soil erosion. Plowing and fertilizers were introduced to deal with these problems-but once again involved longer hours of work and degradation of soil resources(Boserup, The Conditions of Agricultural Growth, Allen and Unwin, 1965, expanded and updated in Population and Technology, Blackwell, 1980.).

While the point of industrial agriculture is lower cost products to create greater productivity thus a higher standard of living as measured by available goods and services, industrial methods have side effects both good and bad. Further, industrial agriculture is not some single indivisible thing, but instead is composed of numerous separate elements, each of which can be modified, and in fact is modified in response to market conditions, government regulation, and scientific advances. So the question then becomes for each specific element that goes into an industrial agriculture method or technique or process: What bad side effects are bad enough that the financial gain and good side effects are outweighed? Different interest groups not only reach different conclusions on this, but also recommend differing solutions, which then become factors in changing both market conditions and government regulations.

Society

The major challenges and issues faced by society concerning industrial agriculture include:

Maximizing the benefits:

- Cheap and plentiful food
- Convenience for the consumer
- The contribution to our economy on many levels, from growers to harvesters to processors to sellers while minimizing the downsides:
- Environmental and social costs
- Damage to fisheries
- Cleanup of surface and groundwater polluted with animal waste
- Increased health risks from pesticides
- Increased ozone pollution and global warming from heavy use of fossil fuels

Benefits

Cheap and Plentiful Food

Population (est.) 10,000 BCE – 2000 CE.

Very roughly:

- 30,000 years ago hunter-gatherer behavior fed 6 million people
- 3,000 years ago primitive agriculture fed 60 million people
- 300 years ago intensive agriculture fed 600 million people
- Today industrial agriculture feeds 6000 million people

Estimated world population at various dates, in thousands

Year	World	Africa	Asia	Europe	Central & South America	North America*	Oceania
8000 BCE	8 000						
1000 BCE	50 000						
500 BCE	100 000						
1 CE	200,000 plus						
1000	310 000						
1750	791 000	106 000	502 000	163 000	16 000	2 000	2 000
1800	978 000	107 000	635 000	203 000	24 000	7 000	2 000
1850	1 262 000	111 000	809 000	276 000	38 000	26 000	2 000
1900	1 650 000	133 000	947 000	408 000	74 000	82 000	6 000
1950	2 518 629	221 214	1 398 488	547 403	167 097	171 616	12 812
1955	2 755 823	246 746	1 541 947	575 184	190 797	186 884	14 265
1960	2 981 659	277 398	1 674 336	601 401	209 303	204 152	15 888
1965	3 334 874	313 744	1 899 424	634 026	250 452	219 570	17 657
1970	3 692 492	357 283	2 143 118	655 855	284 856	231 937	19 443
1975	4 068 109	408 160	2 397 512	675 542	321 906	243 425	21 564
1980	4 434 682	469 618	2 632 335	692 431	361 401	256 068	22 828
1985	4 830 979	541 814	2 887 552	706 009	401 469	269 456	24 678
1990	5 263 593	622 443	3 167 807	721 582	441 525	283 549	26 687
1995	5 674 380	707 462	3 430 052	727 405	481 099	299 438	28 924
2000	6 070 581	795 671	3 679 737	727 986	520 229	315 915	31 043
2005	6 453 628	887 964	3 917 508	724 722	558 281	332 156	32 998**

An example of industrial agriculture providing cheap and plentiful food is the U.S.'s "most successful program of agricultural development of any country in the world". Between 1930 and 2000 U.S. agricultural productivity (output divided by all inputs) rose by an average of about 2 percent annually causing food prices paid by consumers to decrease. "The percentage of U.S. disposable income spent on food prepared at home decreased, from 22 percent as late as 1950 to 7 percent by the end of the century."

Convenience and Choice

Industrial agriculture treats farmed products in terms of minimizing inputs and maximizing outputs at every stage from the natural resources of sun, land and water to the consumer which results in a vertically integrated industry that genetically manipulates crops and livestock; and processes, packages, and markets in whatever way generates maximum return on investment creating convenience foods many customers will pay a premium for. A consumer backlash against food sold for taste, convenience, and profit rather than nutrition and other values (e.g. reduce waste, be natural, be ethical) has led the industry to also provide organic food, minimally processed foods, and minimally packaged foods to maximally satisfy all segments of society thus generating maximum return on investment.

Liabilities

Environment

Industrial agriculture uses huge amounts of water, energy, and industrial chemicals; increasing pollution in the arable land, useable water and atmosphere. Herbicides, insecticides, fertilizers, and animal waste products are accumulating in ground and surface waters. "Many of the negative effects of industrial agriculture are remote from fields and farms. Nitrogen compounds from the Midwest, for example, travel down the Mississippi to degrade coastal fisheries in the Gulf of Mexico. But other adverse effects are showing up within agricultural production systems — for example, the rapidly developing resistance among pests is rendering our arsenal of herbicides and insecticides increasingly ineffective."

Social

A study done for the US. Office of Technology Assessment conducted by the UC Davis Macrosocial Accounting Project concluded that industrial agriculture is associated with substantial deterioration of human living conditions in nearby rural communities.

Animals

"Confined animal feeding operations" or "intensive livestock operations" or "factory farms", can hold large numbers (some up to hundreds of thousands) of animals, often indoors. These animals are typically cows, hogs, turkeys, or chickens. The distinctive characteristics of such farms is the concentration of livestock in a given space. The aim of the operation is to produce as much meat, eggs, or milk at the lowest possible cost.

Food and water is supplied in place, and artificial methods are often employed to maintain animal health and improve production, such as therapeutic use of antimicrobial agents, vitamin supplements and growth hormones. Growth hormones are not used in chicken meat production nor are they used in the European Union for any animal. In meat production, methods are also sometimes employed to control undesirable behaviours often related to stresses of being confined in restricted areas with other animals. More docile breeds are sought (with natural dominant behaviours bred out for example), physical restraints to stop interaction, such as individual cages for chickens, or animals physically modified, such as the de-beaking of chickens to reduce the harm of fighting. Weight gain is encouraged by the provision of plentiful supplies of food to animals breed for weight gain.

The designation "confined animal feeding operation" in the U.S. resulted from that country's 1972 Federal Clean Water Act, which was enacted to protect and restore lakes and rivers to a "fishable, swimmable" quality. The United States Environmental Protection Agency (EPA) identified certain animal feeding operations, along with many other types of industry, as point source polluters of groundwater. These operations were designated as CAFOs and subject to special anti-pollution regulation. In 24 states in the U.S., isolated cases of groundwater contamination has been linked to CAFOs. For example, the ten million hogs in North Carolina generate 19 million tons of waste per year. The U.S. federal government acknowledges the waste disposal issue and requires that animal waste be stored in lagoons. These lagoons can be as large as 7.5 acres (30,000 m). Lagoons not protected with an impermeable liner can leak waste into groundwater under some conditions, as can runoff from manure spread back onto fields as fertilizer in the case of an unforeseen heavy rainfall. A lagoon that burst in 1995 released 25 million gallons of nitrous sludge in North Carolina's New River. The spill allegedly killed eight to ten million fish.

The large concentration of animals, animal waste, and dead animals in a small space poses ethical issues. Animal rights and animal welfare activists have charged that intensive animal rearing is cruel to animals. As they become more common, so do concerns about air pollution and ground water contamination, and the effects on human health of the pollution and the use of antibiotics and growth hormones.

One particular problem with farms on which animals are intensively reared is the growth of antibiotic resistant bacteria. Because large numbers of animals are confined in a small space, any disease would spread quickly, and so antibiotics are used preventively. A small percentage of bacteria are not killed by the drugs, which may infect human beings if it becomes airborne.

According to the U.S. Centers for Disease Control and Prevention (CDC), farms on which animals are intensively reared can cause adverse health reactions in farm workers. Workers may develop acute and chronic lung disease, musculoskeletal injuries, and may catch infections that transmit from animals to human beings.

The CDC writes that chemical, bacterial, and viral compounds from animal waste may travel in the soil and water. Residents near such farms report nuisances such as unpleasant smells and flies, as well as adverse health effects.

The CDC has identified a number of pollutants associated with the discharge of animal waste into rivers and lakes, and into the air. The use of antibiotics may create antibiotic-resistant pathogens; parasites, bacteria, and viruses may be spread; ammonia, nitrogen, and phosphorus can reduce oxygen in surface waters and contaminate drinking water; pesticides and hormones may cause hormone-related changes in fish; animal feed and feathers may stunt the growth of desirable plants in surface waters and provide nutrients to disease-causing micro-organisms; trace elements such as arsenic and copper, which are harmful to human health, may contaminate surface waters.

Crops

The projects within the Green Revolution spread technologies that had already existed, but had not been widely used outside of industrialized nations. These technologies included pesticides, irrigation projects, and synthetic nitrogen fertilizer.

The novel technological development of the Green Revolution was the production of what some referred to as "miracle seeds." Scientists created strains of maize, wheat, and rice that are generally referred to as HYVs or "high-yielding varieties." HYVs have an increased nitrogen-absorbing potential compared to other varieties. Since cereals that absorbed extra nitrogen would typically lodge, or fall over before harvest, semi-dwarfing genes were bred into their genomes. Norin 10 wheat, a variety developed by Orville Vogel from Japanese dwarf wheat varieties, was instrumental in developing Green Revolution wheat cultivars. IR8, the first widely implemented HYV rice to be developed by IRRI, was created through a cross between an Indonesian variety named "Peta" and a Chinese variety named "Dee Geo Woo Gen."

With the availability of molecular genetics in Arabidopsis and rice the mutant genes responsible (*reduced height(rht)*, *gibberellin insensitive (gai1)* and *slender rice (slr1)*) have been cloned and identified as cellular signalling components of gibberellic acid, a phytohormone involved in regulating stem growth via its effect on cell division. Stem growth in the mutant background is significantly reduced leading to the dwarf phenotype. Photosynthetic investment in the stem is reduced dramatically as the shorter plants are inherently more stable mechanically. Assimilates become redirected to grain production, amplifying in particular the effect of chemical fertilisers on commercial yield.

HYVs significantly outperform traditional varieties in the presence of adequate irrigation, pesticides, and fertilizers. In the absence of these inputs, traditional varieties may outperform HYVs. One criticism of HYVs is that they were developed as F1 hybrids, meaning they need to be purchased by a farmer every season rather than saved from previous seasons, thus increasing a farmer's cost of production.

Sustainable Agriculture

The idea and practice of sustainable agriculture has arisen in response to the problems of industrial agriculture. Sustainable agriculture integrates three main goals: environmental stewardship, farm profitability, and prosperous farming communities. These goals have been defined by a variety of disciplines and may be looked at from the vantage point of the farmer or the consumer.

Organic Farming Methods

Organic farming methods combine some aspects of scientific knowledge and highly limited modern technology with traditional farming practices; accepting some of the methods of industrial agriculture while rejecting others. Organic methods rely on naturally occurring biological processes, which often take place over extended periods of time, and a holistic approach; while chemical-based farming focuses on immediate, isolated effects and reductionist strategies.

Integrated Multi-Trophic Aquaculture is an example of this holistic approach. Integrated Multi-Trophic Aquaculture (IMTA) is a practice in which the by-products (wastes) from one species are recycled to become inputs (fertilizers, food) for another. Fed aquaculture (e.g. fish, shrimp) is combined with inorganic extractive (e.g. seaweed) and organic extractive (e.g.

shellfish) aquaculture to create balanced systems for environmental sustainability (biomitigation), economic stability (product diversification and risk reduction) and social acceptability (better management practices).

ORGANIC FARMING

Organic farming is a form of agriculture that relies on crop rotation, green manure, compost, biological pest control, and mechanical cultivation to maintain soil productivity and control pests, excluding or strictly limiting the use of synthetic fertilizers and synthetic pesticides, plant growth regulators, livestock feed additives, and genetically modified organisms. Since 1990 the market for organic products has grown at a rapid pace, averaging 20-25 percent per year to reach $33 billion in 2005. This demand has driven a similar increase in organically managed farmland. Approximately 306,000 square kilometres (30.6 million hectares) worldwide are now farmed organically, representing approximately 2% of total world farmland. In addition, as of 2005 organic wild products are farmed on approximately 62 million hectares. (IFOAM 2007:10).

Organic agricultural methods are internationally regulated and legally enforced by many nations, based in large part on the standards set by the International Federation of Organic Agriculture Movements (IFOAM), an international umbrella organization for organic organizations established in 1972. IFOAM defines the overarching goal of organic farming as follows:

"Organic agriculture is a production system that sustains the health of soils, ecosystems and people. It relies on ecological processes, biodiversity and cycles adapted to local conditions, rather than the use of inputs with adverse effects. Organic agriculture combines tradition, innovation and science to benefit the shared environment and promote fair relationships and a good quality of life for all involved.." —International Federation of Organic Agriculture Movements

The organic movement began in the early 1930s and early 1940s as a reaction to agriculture's growing reliance on synthetic fertilizers. Artificial fertilizers had been created during the

18th century, initially with superphosphates and then ammonia derived fertilizers mass-produced using the Haber-Bosch process developed during World War I. These early fertilizers were cheap, powerful, and easy to transport in bulk. Similar advances occurred in chemical pesticides in the 1940s, leading to the decade being referred to as the 'pesticide era'.

Sir Albert Howard is widely considered to be the father of organic farming. Further work was done by J.I. Rodale in the United States, Lady Eve Balfour in the United Kingdom, and many others across the world.

As a percentage of total agricultural output, organic farming has remained tiny since its beginning. As environmental awareness and concern increased amongst the general population, the originally supply-driven movement became demand-driven. Premium prices from consumers and in some cases government subsidies attracted many farmers into converting. In the developing world, many farmers farm according to traditional methods which are comparable to organic farming but are not certified. In other cases, farmers in the developing world have converted out of necessity. As a proportion of total global agricultural output, organic output remains small, but it has been growing rapidly in many countries, notably in Europe.

Methods

"An organic farm, properly speaking, is not one that uses certain methods and substances and avoids others; it is a farm whose structure is formed in imitation of the structure of a natural system that has the integrity, the independence and the benign dependence of an organism" —Wendell Berry, "The Gift of Good Land"

The term holistic is often used to describe organic farming. Enhancing soil health is the cornerstone of organic farming. A variety of methods are employed, including crop rotation, green manure, cover cropping, application of compost, and mulching. Organic farmers also use certain processed fertilizers such as seed meal, and various mineral powders such as rock phosphate and greensand, a naturally occurring form of potash. These methods help to control erosion, promote biodiversity, and enhance the health of the soil.

Pest control targets animal pests (including insects), fungi,weeds and disease. Organic pest control involves the cumulative effect of many techniques, including, allowing for an acceptable level of pest damage, encouraging or even introducing beneficial organisms, careful crop selection and crop rotation, and mechanical controls such as row covers and traps. These techniques generally provide benefits in addition to pest control—soil protection and improvement, fertilization, pollination, water conservation, season extension, etc.—and these benefits are both complementary and cumulative in overall effect on farm health. Effective organic pest control requires a thorough understanding of pest life cycles and interactions.

Most organic farms use less pesticides than most conventional farms. The main three used are Bt (a bacterial toxin), pyrethrum and rotenone. Surveys have found that fewer than 10% of organic farmers use these pesticides regularly; one survey found that only 5.3% of vegetable growers in California use rotenone while 1.7% use pyrethrum (Lotter 2003:26).

Weeds are controlled mechanically, thermically and through the use of covercrops and mulches. The traditional method is to remove weeds by hand, which is still practiced in developing countries by small scale farmers. However, this has proven too costly in developed countries where labor is more expensive. One recent innovation in rice farming is to introduce ducks and fish to wet paddy fields, which eat both weeds and insects.

Standards

Standards regulate production methods and in some cases final output for organic agriculture. Standards may be voluntary or legislated. As early as the 1970s organic producers could be voluntarily certified by private associations. In the 1980s, governments began to produce organic production guidelines. Beginning in the 1990s, a trend toward legislation of standards began, most notably with the 1991 EU-Eco-regulation developed for European Union, which set standards for 12 countries, and a 1993 UK program. The EU's program was followed by a

Japan program in 2001, and in 2002 the United States created the National Organic Program (NOP). As of 2007 over 60 countries have regulations on organic farming (IFOAM 2007:11). In 2005 IFOAM created the Principles of Organic Agriculture, an international guideline for certification criteria. Typically the agencies do not certify individual farms, but rather accredit certification groups.

Materials used in organic production and foods are tested independently by the Organic Materials Review Institute.

Composting

Under USDA organic standards, manure must be composted and allowed to reach a sterilizing temperature. If raw animal manure is used, 120 days must pass before the crop is harvested.

Economics

The economics of organic farming, a subfield of agricultural economics, encompasses the entire process and effects of organic farming in terms of human society, including social costs, opportunity costs, unintended consequences, information asymmetries, and economies of scale. Although the scope of economics is broad, agricultural economics tends to focus on maximizing yields and efficiency at the farm level.

Mainstream economics takes an anthropocentric approach to the value of the natural world: biodiversity, for example, is considered beneficial only to the extent that it is valued by people and increases profits. Some governments such as the European Union subsidize organic farming, in large part because these countries believe in the external benefits of reduced water use, reduced water contamination by pesticides and nutrients of organic farming, reduced soil erosion, reduced carbon emissions, increased biodiversity, and assorted other benefits.

Organic farming is labor and knowledge-intensive whereas conventional farming is capital-intensive, requiring more energy and manufactured inputs. Organic farmers in California have cited marketing as their greatest obstacle.

Geographic Producer Distribution

The markets for organic products are strongest in North America and Europe, which as of 2001 are estimated to have $6 and $8 billion respectively of the $20 billion market (2003:6). However, as of 2007 organic farmland is distributed across the globe. Australasia has 39% of the total organic farmland with Australia's 11.8 million hectares, but 97 percent of this land is sprawling rangeland (2007:35), which results in total sales of approximately 5% of US sales (2003:7). Europe has 23 percent of total organic farmland (6.9 million hectares), followed by Latin America with 19 percent (5.8 million hectares). Asia has 9.5 percent while North America has 7.2 percent. Africa has a mere 3 percent. See also Organic farming by country.

Besides Australia, the countries with the most organic area are Argentina (3.1 million hectares), China (2.3 million hectares), and the United States (1.6 million hectares). Much of Argentina's organic farmland is pasture, like that of Australia (2007:42). Italy, Spain, Germany, Brazil, Uruguay, and the UK follow the United States by the amount of land managed organically (2007:26).

Growth

As of 2001, the estimated total market value of certified organic products was estimated to be $20 billion. By 2002 this was $23 billion and by 2005 $33 billion, with Organic Monitor projecting sales of $40 billion in 2006 (IFOAM 2007:11). The change from 2001 to 2005 represents a compound growth of 10.6 percent.

In recent years both Europe and North America have experienced strong growth in organic farmland. Each added half a million hectares from 2004 to 2007 — for the US this is a 29 percent change (IFOAM 2007:11,27). However, this growth has occurred under different conditions. While the European Union has shifted agricultural subsidies to organic farmers in recognition of its environmental benefits, the United States has taken a free market approach. As a result, as of 2001 3 percent of European farmland was organically managed compared to just 0.3 percent of United States farmland (Lotter

2003:7). By 2005 Europe's organic land was 3.9 percent while the United States' had risen to 0.6 percent (IFOAM 2007:14-15).

IFOAM's The World of Organic Agriculture: Statistics and Emerging Trends 2007 lists the countries which added the most hectares and had the highest percentage growth in 2007 (IFOAM 2007:27-28). Among these, China is listed third in adding the most hectares behind the United States and Argentina. China jumped from approximately 300,000 hectares of organic land in 2005 to approximately 3.5 million hectares in 2006 — an increase of over a thousand percent. This rise can be attributed to the certification of China's Organic Food Development Center in 2002 by IFOAM. The end of 2005 marks the end of the three-year transition period begun in 2002..

Productivity and Profitability

A 2006 study suggests that converted organic farms have lower pre-harvest yields than their conventional counterparts in developed countries (92%) and that organic farms have higher pre-harvest yields than their low-intensity counterparts in developing countries (132%). The researcher attributes this to a relative lack of expensive fertilizers and pesticides in the developing world compared to the intensive, subsidy-driven farming of the developed world. Nonetheless, the researcher purposely avoids making the claim that organic methods routinely outperform green-revolution (conventional) methods. This study incorporated a 1990 review of 205 crop comparisons which found that organic crops had 91% of conventional yields. A major US survey published in 2001, analyzed results from 150 growing seasons for various crops and concluded that organic yields were 95-100% of conventional yields.

Lotter (2003:10) reports that repeated studies have found that organic farms withstand severe weather conditions better than conventional farms, sometimes yielding 70-90% more than conventional farms during droughts. A 22-year farm trial study by Cornell University published in 2005 concluded that organic farming produces the same corn and soybean yields as conventional methods over the long-term averages, but consumed less energy and used zero pesticides. The results were attributed to lower yields in general but higher yields during drought years. A study of 1,804 organic farms in Central American hit by Hurricane Mitch in 1998 found that the organic farms sustained the damage much better, retaining 20 to 40% more topsoil and smaller economic losses at highly significant levels than their neighbors.

On the other hand, a prominent 21-year Swiss study found an average of 20% lower organic yields over conventional, along with 50% lower expenditure on fertilizer and energy, and 97% less pesticides. A long-term study by U.S Department of Agriculture Agricultural Research Service (ARS) scientists concluded that, contrary to widespread belief, organic farming can build up soil organic matter better than conventional no-till farming, which suggests long-term yield benefits from organic farming. An 18-year study of organic methods on nutrient-depleted soil concluded that conventional methods were superior for soil fertility and yield in a cold-temperate climate, arguing that much of the benefits from organic farming are derived from imported materials which could not be regarded as "self-sustaining".

While organic farms have lower yields, organic methods require no synthetic fertilizer and pesticides. The decreased cost on those inputs, along with the premiums which consumers pay for organic produce, create higher profits for organic farmers. Organic farms have been consistently found to be as or more profitable than conventional farms with premiums included, but without premiums profitability is mixed (Lotter 2003:11). Welsh (1999) reports that organic farmers are more profitable in the drier states of the United States, likely due to their superior drought performance.

In 2008 the UN Environmental Programme (UNEP) and UN Conference on Trade and Development (UNCTAD) issued a report which stated that "organic agriculture can be more conducive to food security in Africa than most conventional production systems, and that it is more likely to be sustainable in the long-term". The report assessed 114 projects in 24 African countries, finding that "yields had more than doubled where organic, or near-organic practices had been used" and that soil fertility and drought resistance improved.

Macroeconomic Impact

Organic methods often require more labor, providing rural jobs but increasing costs to urban consumers.

Motivations

Agriculture in general imposes external costs upon society through pesticides, nutrient runoff, excessive water usage, and assorted other problems. As organic methods minimize some of these factors, organic farming is believed to impose fewer external costs upon society. A 2000 assessment of agriculture in the UK determined total external costs costs for 1996 of 2343 million British pounds or 208 pounds per hectare. A 2005 analysis of these costs in the USA concluded that cropland imposes approximately 5 to 16 billion dollars ($30 to $96 per hectare), while livestock production imposes 714 million dollars. Both studies concluded that more should be done to internalize external costs, and neither included subsidies in their analysis, but noted that subsidies also influence the cost of agriculture to society. Both focused on purely fiscal impacts. The 2000 review included reported pesticide poisonings but did not include speculative chronic effects of pesticides, and the 2004 review relied on a 1992 estimate of the total impact of pesticides.

Pesticides

Some pesticides may damage human health and the environment, and most organic farms use less pesticides than conventional farms. The main three pesticides used in organic farming are Bt (a bacterial toxin), pyrethrum, rotenone, copper and sulphur. Surveys have found that fewer than 10% of organic farmers use these pesticides regularly; one survey found that only 5.3% of vegetable growers in California use rotenone while 1.7% use pyrethrum (Lotter 2003:26). Reduction and elimination of chemical pesticide use is technologically challenging. Few organic farms manage to eliminate the use of pesticides entirely; organic pesticides are often used to compliment other pest control strategies.

Pesticide runoff is one of the most significant effects of pesticide use. The USDA Natural Resources Conservation Service tracks the environmental risk posed by pesticide water contamination from farms, and its conclusion has been that "the Nation's pesticide policies during the last twenty six years have succeeded in reducing overall environmental risk, in spite of slight increases in area planted and weight of pesticides applied. Nevertheless, there are still areas of the country where there is no evidence of progress, and areas where risk levels for protection of drinking water, fish, algae and crustaceans remain high".

Pest resistant genetically modified crops have been proposed as an alternative to pesticide use, however concerns over the safety and the long term benefits of genetically modified food, result in the genetic modification being widely opposed in the organic farming movement.

Food Quality

Organic food is widely believed by the lay public to be healthier, although the research is controversial. Two studies have found that children fed organic diets experienced significantly lower organophosphorus pesticide exposure than children fed conventional diets. Although the researchers did not collect health *outcome* data in this study, they concluded "it is intuitive to assume that children whose diets consist of organic food items would have a lower probability of neurologic health risks". A 2007 study found that consumption of organic milk is associated with a decrease in risk for eczema, although no comparable benefit was found for organic fruits, vegetables, or meat.

Extensive scientific research is being carried out in Switzerland at over 200 farms to determine differences in the quality of organic food products compared to conventional in addition to other tests. The FiBL Institute has been investigating the differences at over 200 farms. It states that "organic products stand out as having higher levels of secondary plant compounds and vitamin C. In the case of milk and meat, the fatty acid profile is often better from a nutritional point of view. As far as carbohydrates and minerals, organic products are no different from conventional products. However, in regard to undesirables such as nitrate and pesticide residues, organic products have a clear advantage. A £12m EU-funded investigation into the difference between organic and ordinary farming published in 2007 found that organic foods have more nutritional value. A recent study found that organically grown produce has double the flavonoids, an important antioxidant.. A 2007 study found that organically grown kiwifruit had more antioxidants than conventional kiwifruit.

Genetically Modified Organisms

A key characteristic of organic farming is rejection of genetically engineered products, including plants and animals. On October 19, 1998, participants at IFOAM's 12th Scientific Conference of IFOAM) issued the Mar del Plata Declaration, where more than 600 delegates from over 60 countries voted unanimously to exclude the use of genetically modified organisms in food production and agriculture. From this point, it became widely recognized that GMOs are categorically excluded from organic farming. Although opposition to the use of any transgenic technologies in organic farming is strong, agricultural researchers Luis Herrera-Estrella & Ariel Alvarez-Morales continue to advocate integration of transgenic technologies into organic farming as the optimal means to sustainable agriculture, particularly in the developing world. Similarly, some organic farmers question the rationale behind the ban on the use of genetically engineered seed because they see it a biological technology consistent with organic principles

Although GMOs are excluded from use in organic farming, there is concern that the pollen from genetically modified crops is increasingly contaminating organic and heirloom genetics making it difficult, if not impossible, to keep these genetics from entering the organic food supply. International trade restrictions limit the availability GMOs to certain countries.

The actual dangers that genetic modification could pose to the environment or, supposedly, individual health, are hotly contended. See GM food controversy.

Soil Conservation

In *Dirt: The Erosion of Civilizations*, geomorphologist David Montgomery outlines a coming crisis from soil erosion. Agriculture relies on roughly one meter of topsoil, and that is being depleted ten times faster than it is being replaced. No-till farming, which some claim depends upon pesticides, is regarded as one way to minimize erosion. However, a recent study by the USDA's Agricultural Research Service has found that organic farming is even better at building up the soil than no-till.

Cite error: Closing </ref> missing for <ref> tag According to the Rodale Institute, which has been comparing organic agricultural systems and conventional systems since 1981, organic agriculture also can be used to mitigate global warming by decreasing fossil fuel emissions and sequestering carbon in the soil. The elimination of synthetic nitrogen in organic systems decreases fossil fuel consumption by 33 percent (LaSalle)and carbon sequestration takes CO_2 out of the atmosphere by putting it in the soil in the form of organic matter which is often lost in conventionally managed soils. Carbon sequestration occurs at especially high levels in organic no-till managed soil according to the Rodale Institute.

Nutrient Leaching

Excess nutrients in lakes, rivers, and groundwater can cause algal blooms, eutrophication, and subsequent dead zones. In addition, nitrates are harmful to aquatic organisms by themselves. The main contributor to this pollution is nitrate fertilizers whose use is expected to "double or almost triple by 2050". Researchers at the United States National Academy of Sciences found that that organically fertilizing fields "significantly [reduces] harmful nitrate leaching" over conventionally fertilized fields: "annual nitrate leaching was 4.4-5.6 times higher in conventional plots than organic plots".

Scientists believe that the large dead zone in the Gulf of Mexico is caused in large part by agricultural pollution: a combination of fertilizer runoff and livestock manure runoff. A study by the United States Geological Survey (USGS) found that over half of the nitrogen released into the Gulf comes from agriculture. The economic cost of this for fishermen may be large, as they must travel far from the coast to find fish.

At the 2000 IFOAM Conference, researchers presented a study of nitrogen leaching into the Danube River. They found that nitrogen runoff was substantially lower among organic farms and suggested that the external cost could be internalized by charging 1 euro per kg of nitrogen released.

A 2005 study found a strong link between agricultural runoff and algae blooms in California.

Biodiversity

A wide range of organisms benefit from organic farming, but it is unclear whether organic methods confer greater benefits than integrated agri-environmental conventional programs. Nearly all non-crop, naturally-occurring species observed in comparative farm land practice studies show a preference in organic farming both by population and richness. Spanning all associated species, there is an average of 30% more on organic farms versus conventional farming methods. Birds, butterflies, soil microbes, beetles, earthworms, spiders, vegetation, and mammals are particularly affected.

Organic crops use little or no herbicides and pesticides and thus biodiversity fitness and population density benefit. Many weed species attract beneficial insects that improve soil qualities and forage on weed pests. Soil-bound organisms often benefit because of increased bacteria populations due to natural fertilizer spread such as manure, while experiencing reduced intake of herbicides and pesticides commonly associated with conventional farming methods. Increased biodiversity, especially from soil microbes such as mycorhizzae, have been proposed as an explanation for the high yields experienced by some organic plots, especially in light of the differences seen in a 21-year comparison of organic and control fields.

The level of biodiversity that can be yielded from organic farming provides a natural capital to humans. Species found in most organic farms provides a means of agricultural sustainability by reducing amount of human input (e.g. fertilizers, pesticides). Farmers that produce with organic methods reduce risk of poor yields by promoting biodiversity. Common game birds such as the ring-necked pheasant and the northern bobwhite often reside in agriculture landscapes, and are a natural capital yielded from high demands of recreational hunting. Because bird species richness and population are typically higher on organic farm systems, promoting biodiversity can be seen as logical and economical.

Biological research on soil and soil organisms has proven beneficial to the system of organic farming. Varieties of bacteria and fungi break down chemicals, plant matter and animal waste into productive soil nutrients. In turn, the producer benefits by healthier yields and more arable soil for future crops.

Furthermore, a 21-year study was conducted testing the effects of organic soil matter and its relationship to soil quality and yield. Controls included actively managed soil with varying levels of manure, compared to a plot with no manure input. After the study commenced, there was significantly lower yields on the control plot when compared to the fields with manure. The concluded reason was an increased soil microbe community in the manure fields, providing a healthier, more arable soil system.

Sales and Marketing

Organic farmers report that marketing and distribution are difficult obstacles. Most of organic sales are concentrated in developed nations. These products are what economists call credence goods in that they rely on uncertain certification. As food prices rise, organic products may experience falling demand. A 2008 survey by WSL Strategic Retail found that interest in organic products had dropped since 2006, and that 42% of Americans polled don't trust organic produce. and The Hartman Group reports that 69% of Americans claim to occasionally buy organic products, down from 73% in 2005. The Hartman Group says that people may be substituting local produce for organic produce.

Distributors

In the United States, 75% of organic farms are smaller than 2.5 hectares and in California 2% of the farms account for over half of the sales (Lotter 2003:4). Groups of small farms join together in cooperatives such as Organic Valley, Inc. to market their goods more effectively.

Over the past twenty years, however, most of these cooperative distributors have merged or been bought out. Rural sociologist Philip H. Howard has researched the structure and transformation of the organic industry in the United States. He claims that in 1982 there were 28 consumer cooperative distributors but as of 2007 there are only 3, and he has created a graphic displaying the consolidation. His research shows that most of these small cooperatives have been absorbed into large multinational corporations such as General Mills, Heinz, ConAgra, Kellog, and assorted other brands. This consolidation has raised concerns among consumers and journalists of potential fraud and degradation in standards. Most of these large corporations sell their organic products through subsidiaries, allowing them to keep their names off the labels.

Farmers' Markets

Price premiums are important for the profitability of small organic farmers, and so many sell directly to consumers in farmers' markets. In the United States the number of farmers' markets has grown from 1,755 in 1994 to 4,385 in 2006.

Capacity Building

Organic agriculture can contribute to meaningful socio-economic and ecologically sustainable development, especially in poorer countries. On one hand, this is due to the application of organic principles, which means efficient management of local resources (e.g. local seed varieties, manure, etc.) and therefore cost-effectiveness. On the other hand, the market for organic products – at local and international level – has tremendous growth prospects and offers creative producers and exporters in the South excellent opportunities to improve their income and living conditions.

Organic Agriculture is a very knowledge intensive production system. Therefore capacity building efforts play a central role in this regard. There are many efforts all around the world regarding the development of training material and the organization of training courses related to Organic Agriculture. Big parts of existing knowledge is still scattered and not easy accessible. Especially in Developing Countries this situation remains an important constraint for the growth of the organic sector.

For that reason, the International Federation of Organic Agriculture Movements created an Internet Training Platform whose objective is to become the global reference point for Organic Agriculture training through free access to high quality training materials and training programs on Organic Agriculture. In November 2007, the Training Platform hosted more than 170 free manuals and 75 training opportunities.

Controversy

A number of critics contest the notion that organic agricultural systems are more friendly to the environment and more sustainable than high-yielding farming systems. Among these critics are Norman Borlaug, father of the "green revolution," and winner of the Nobel Peace Prize, who asserts that organic farming practices can at most feed 4 billion people, after expanding cropland dramatically and destroying ecosystems in the process and Prof A. Trewavas.

The debate has been summarized in an exchange between Trewavas and Lord P. Melchett, and published by a major supermarket, concerned about examining the issues.

One study from the Danish Environmental Protection Agency found that, area-for-area, organic farms of potatoes, sugar beet and seed grass produce as little as half the output of conventional farming.

In 2008 a study from UN Environmental Programme concluded that organic methods greatly increase yields in Africa and a review of over two hundred crop comparisons argued that organic farming could produce enough food per capita to sustain the current human population; the difference in yields between organic and non-organic methods were small, with non-organic methods resulting in slightly higher yields in developed areas and organic methods resulting in slightly higher yields in developing areas.

That analysis has been severely criticised by Alex Avery, who contends that the review claimed many non-organic studies to be organic, misreported organic yields, made false comparisons between yields of organic and non-organic studies which were not comparable, counted high organic yields several times by citing different papers which referenced the same data, and gave equal weight to studies from sources which were not impartial and rigorous university studies.

Urs Niggli, director of the FiBL Institute contents that the wave of newspaper articles like 'Organic food exposed' or 'The hypocrisy of organic farmers' are a part of a global campaign against organic farming that take their arguments mostly from the book 'The truth about organic farming', by Alex Avery of the Hudson Institute.

In 1998, Dennis Avery of the Hudson Institute claimed the risk of E. coli infection was eight times higher when eating organic food rather than non-organic food, using the Center for Disease Control (CDC) as a source. When the CDC was contacted, it stated that there was no evidence for the claim. The *New York Times* commented on Avery's attacks: "The attack on organic food by a well-financed research organization suggests that, though organic food accounts for only 1 percent of food sales in the United States, the conventional food industry is worried."

Organic agriculture can reduce the level of negative externalities from (conventional) agriculture. Whether this is seen as private or public benefits depends upon the initial specification of property rights. However, it is clear that agriculture has been undervalued and underestimated as a means to combat global climate change. Soil carbon data recorded by The Rodale Institute show that regenerative organic agricultural practices are among the most effective strategies for mitigating CO_2 emissions.

PERMACULTURE

Permaculture is an approach to designing human settlements and perennial agricultural systems that mimic the relationships found in the natural ecologies. It was first developed by Australians Bill Mollison and David Holmgren and their associates during the 1970s in a series of publications. The word *permaculture* is a portmanteau of *permanent agriculture*, as well as *permanent culture*.

Permaculture design principles extend from the position that "The only ethical decision is to take responsibility for our own existence and that of our children" (Mollison, 1990). The intent was that, by rapidly training individuals in a core set of design principles, those individuals could design their own environments and build increasingly self-sufficient human settlements — ones that reduce society's reliance on industrial systems of production and distribution that Mollison identified as fundamentally and systematically destroying Earth's ecosystems.

While originating as an agro-ecological design theory, permaculture has developed a large international following of individuals who have received training through intensive two week long 'permaculture design courses'. This 'permaculture community' continues to expand on the original ideas, integrating a range of ideas of alternative culture, through a network of training, publications, permaculture gardens, and internet forums. In this way, permaculture has become both a design system and a loosely defined philosophy or lifestyle ethic.

History

The term *permanent agriculture* was coined by Franklin Hiram King in his classic book from 1911, *Farmers of Forty Centuries: Or Permanent Agriculture in China, Korea and Japan*. In this context, permanent agriculture is understood as agriculture that can be sustained indefinitely.

This definition was supported by Australian P. A. Yeomans (*Water for Every Farm*, 1973) who introduced an observation-based approach to land use in Australia in the 1940s, based partially on his understanding of geology. Yeomans introduced Keyline Design as a way of managing the supply and distribution of water of a site. Holmgren based his EcoVillage design on the keyline principle, (see WikiMapia view)

The work of Howard T. Odum was also an early influence, especially for Holmgren. Odum's work focused on system ecology, in particular the Maximum power principle, which examines the energy of a system and how natural systems tend to maximise the energy embodied in a system. For example, the total calorific value of woodland is very high with its multitude of plants and animals. It is an efficient converter of sunlight into biomass. A wheat field, on the other hand, has much less total energy and often requires a large energy input in terms of fertiliser. Another early influence was the work of Esther Deans, who pioneered No-Dig Gardening methods. Other recent influences include the VAC system in Vietnam which is a government supported system to build Vegetable Aquaculture and Animal enClosures that cycle resources.

In the mid 1970s, Australians Bill Mollison and David Holmgren started to develop ideas that they hoped could be used to create stable agricultural systems. This was a result of their perception of a rapidly growing use of destructive industrial-agricultural methods. They saw that these methods were poisoning the land and water, reducing biodiversity, and removing billions of tons of soil from previously fertile landscapes. A design approach called "permaculture" was their response and was first made public with the publication of *Permaculture One* in 1978.

The term *permaculture* initially meant "permanent agriculture" but was quickly expanded to also stand for "permanent culture" as it was seen that social aspects were an integral part of a truly sustainable system. Mollison and Holmgren are widely considered to be the co-originators of the modern permaculture concept.

After the publication of *Permaculture One*, Mollison and Holmgren further refined and developed their ideas by designing hundreds of permaculture sites and organizing this information into more detailed books. Mollison lectured in over 80 countries and his two-week Design Course was taught to many hundreds of students. By the early 1980s, the concept had moved on from being predominantly about the design of agricultural systems towards being a more fully holistic design process for creating sustainable human habitats.

By the mid 1980s, many of the students had become successful practitioners and had themselves begun teaching the techniques they had learned. In a short period of time permaculture groups, projects, associations, and institutes were established in over one hundred countries. In 1991 a four-part Television documentary by ABC productions called 'The Global Gardener' showed permaculture applied to a range of worldwide situations, bringing the concept to a much broader public. Excerpts are available online through YouTube. Permaculture has developed from its origins in Australia into an international 'movement'. English permaculture teacher Patrick Whitefield, author of *The Earth Care Manual* and *Permaculture in a Nutshell*, suggests that there are now two strands of permaculture: a) Original and b) Design permaculture. Original permaculture attempts to closely replicate nature by developing edible ecosystems which closely resemble their wild counterparts. Design permaculture takes the working connections at use in an ecosystem and uses them as its basis. The end result may not look as "natural" as a forest garden, but still has an underlying design based on ecological principles. Through close observation of natural energies and flow patterns efficient design systems can be developed. This has become known as Natural Systems Design. (Dr. M Millington and A Sampson-Kelly)

Elements of Design

Permaculture principles draw heavily on the practical application of ecological theory to analyze the characteristics and potential relationships between design elements. Each element of a design is carefully analyzed in terms of its needs, outputs, and properties. For example a chicken needs water, moderated microclimate, food and other chickens, and produces meat, eggs, feathers and manure and can help break the soil. Design elements are then assembled in relation to one another so that the products of one element feed the needs of adjacent elements. Synergy between design elements is achieved while minimizing waste and the demand for human labour or energy. Exemplary permaculture designs evolve over time, and can become extremely complex mosaics of conventional and inventive cultural systems that produce a high density of food and materials with minimal input.

While techniques and cultural systems are freely borrowed from organic agriculture, sustainable forestry, horticulture, agroforestry, and the land management systems of indigenous peoples, permaculture's fundamental contribution to the field of ecological design is the development of a concise set of broadly applicable organizing principles that can be transferred through a brief intensive training.

Modern Permaculture

Modern permaculture is a system design tool. It is a way of:

1. looking at a whole system or problem;
2. observing how the parts relate;
3. planning to mend sick systems by applying ideas learnt from long-term sustainable working systems;
4. seeing connections between key parts.

In permaculture, practitioners learn from the working systems of nature to plan to fix the damaged landscapes of human agricultural and city systems. This thinking applies to the design of a kitchen tool as easily to the re-design of a farm. Permaculture practitioners apply it to everything deemed necessary to build a sustainable future. Commonly, "Initiatives... tend to evolve from strategies that focus on efficiency (for example, more accurate and controlled uses of inputs and minimisation of waste) to substitution (for example, from more to less disruptive interventions, such as from biocides to more specific biological controls and other more benign alternatives) to redesign (fundamental changes in the design and management of the operation) (Hill & MacRae 1995, Hill et al. 1999)." "Permaculture is about helping people make redesign choices: setting new goals and a shift in thinking that affects not only their home but their actions in the workplace, borrowings and investments" (A Sampson-Kelly and Michel Fanton 1991). Examples include the design and employment of complex transport solutions, optimum use of natural resources such as sunlight, and "radical design of information-rich, multi-storey polyculture systems" (Mollison & Slay 1991).

"This progression generally involves a shift in the nature of one's dependence — from relying primarily on universal, purchased, imported, technology-based interventions to more specific locally available knowledge and skill-based ones. This usually eventually also involves fundamental shifts in world-views, senses of meaning, and associated lifestyles (Hill 1991)." "My experience is that although efficiency and substitution initiatives can make significant contributions to sustainability over the short term, much greater longer-term improvements can only be achieved by redesign strategies; and, furthermore, that steps need to be taken at the outset to ensure that efficiency and substitution strategies can serve as stepping stones and not barriers to redesign..." (Hill 2000)

Core Values

Permaculture is a broad-based and holistic approach that has many applications to all aspects of life. At the heart of permaculture design and practice is a fundamental set of 'core values' or ethics which remain constant whatever a person's situation, whether they are creating systems for town planning or trade; whether the land they care for is only a windowbox or an entire forest. These 'ethics' are often summarised as;

- Earthcare – recognising that Earth is the source of all life (and is possibly itself a living entity — see Gaia theory), that Earth is our valuable home, and that we are a part of Earth, not apart from it.
- Peoplecare – supporting and helping each other to change to ways of living that do not harm ourselves or the planet, and to develop healthy societies.

- Fairshare (or placing limits on consumption)-ensuring that Earth's limited resources are used in ways that are equitable and wise.

Modern thought about permaculture began with the issue of sustainable food production. It started with the belief that for people to feed themselves sustainably, they need to move away from reliance on industrialised agriculture. Where industrial farms use technology powered by fossil fuels (such as gasoline, diesel and natural gas), and each farm specialises in producing high yields of a single crop, permaculture stresses the value of low inputs and diverse crops. The model for this was an abundance of small-scale market and home gardens for food production, and a main issue was food miles.

Design Innovation

The core of permaculture has always been in supplying a design toolkit for human habitation. This toolkit helps the designer to model a final design based on an observation of how ecosystems interact. A simple example of this is how the Sun interacts with a plant by providing it with energy to grow. This plant may then be pollinated by bees or eaten by deer. These may disperse seed to allow other plants to grow into tall trees and provide shelter to these creatures from the wind. The bees may provide food for birds and the trees provide roosting for them. The tree's leaves fall and rot, providing food for small insects and fungus. Such a web of intricate connections allows a diverse population of plant life and animals to survive by giving them food and shelter. One of the innovations of permaculture design was to appreciate the efficiency and productivity of natural ecosystems, to use natural energies (wind, gravity, solar, fire, wave and more) and seek to apply this to the way human needs for food and shelter are met. One of the most notable proponents of this design system has been David Holmgren, who based much of his permaculture innovation on zone analysis.

O'BREDIM Design Methodology

O'BREDIM is a mnemonic and acronym for *observation, boundaries, resources, evaluation, design, implementation and maintenance.*

- Observation allows you first to see how the site functions within itself, to gain an understanding of its initial relationships. Some people recommend a year-long observation of a site before anything is planted. During this period all factors, such as lay of the land, natural flora and so forth, can be brought into the design. A year allows the site to be observed through all seasons, although it must be realised that, particularly in temperate climates, there can be substantial variations between years.
- Boundaries refer to physical ones as well as to those your neighbours might place on you, for example.
- Resources include the people involved, funding, as well as what you can grow or produce in the future.
- Evaluation of the first three will then allow you to prepare for the next three. This is a careful phase of taking stock of what you have at hand to work with.
- Design is a creative and intensive process, and you must stretch your ability to see possible future synergetic relationships.
- Implementation is literally the ground-breaking part of the process when you carefully dig and shape the site.
- Maintenance is then required to keep your site at a healthy optimum, making minor adjustments as necessary. Good design will preclude the need for any major adjustment.

Patterns

The use of patterns both in nature and reusable patterns from other sites is often key to permaculture design. This echoes the Pattern language of Christopher Alexander used in architecture which has been an inspiration for many permaculture designers. All things, even the wind, the waves and the earth on its axis, moving around the Sun, form patterns. In pattern application, permaculture designers are encouraged to develop:

1. Awareness of the patterns that exist in nature (and how these function)

2. Application of pattern on sites in order to satisfy specific design needs.

"The application of pattern on a design site involves the designer recognising the shape and potential to fit these patterns or combinations of patterns comfortably onto the landscape" Sampson-Kelly. Branching can be used for the direction of paths, rather than straight paths with square angles. Lobe-like paths of the main path (known as keyhole paths) can be used to minimise waste and compaction of the soil.

Zones

Permaculture zones are a way of organising design elements in a human environment based on the frequency of human use and plant or animal needs. Frequently manipulated or harvested elements of the design are located close to the house in zones one and two such as herbs for the kitchen, whereas chickens like to be close but need to be kept at a safe distance to reduce noise and contamination (unless they are house trained). Less frequently used or manipulated elements, and elements that benefit from isolation (such as wild species) are farther away.

Links and Connections

Also key to the permacultural design model is that useful connections are made between components in the final design. The formal analogy for this is a natural mature ecosystem. So, in much the same way as there are useful connections between Sun, plants, insects and soil there will be useful connections between different plants and their relationship to the landscape and humans.

Another innovation of the permaculture design is to design a landuse or other system that has multiple outputs. In terms of Holmgren's application of H.T. Odum's work, a useful connection is viewed as one that maximises power: that is, maximizes the rate of useful energy transformation. A comparison which illustrates this is between a wheat field and a forest. "It is not the number of diverse things in a design that leads to stability, it is the number of beneficial connections between these components" Mollison 1988.

Layers/'stacking'

In permaculture and forest gardening, seven layers are identified:

1. The canopy
2. Low tree layer (dwarf fruit trees)
3. Shrubs
4. Herbaceous
5. Rhizosphere (root crops)
6. Soil Surface (cover crops)
7. Vertical Layer (climbers, vines)

An eighth layer, Mycosphere (fungi), is often included.

A mature ecosystem such as ancient woodland has a huge number of relationships between its component parts: trees, understory, ground cover, soil, fungi, insects and other animals. Plants grow at different heights. This allows a diverse community of life to grow in a relatively small space. Plants come into leaf and fruit at different times of year.

Floor in the time before the top canopy re-appears with the spring. A wood suffers very little soil erosion, as there are always roots in the soil. It offers a habitat to a wide variety of animal life, which the plants rely on for pollination and seed distribution. The productivity of such a forest, in terms of how much new growth it produces, exceeds that of the most productive wheat field. It is in this observation-of how much more productive a wood may be on far less fertilizer input-that the potential productivity of a permaculture design is modelled. The many connections in a wood contribute together to a proliferation of opportunities for amplifier feedbacks to evolve that in turn maximise energy flow through the system.

Here is a photo of a layered warm temperate garden in NSW, Australia (courtesy of PermacualtureVisions). There are several layers: the canopy layer is Inga Edulis (ice cream bean), the middle stratum contains plum and peach, mango, mulberry and nurse plants such as native wattle. There are shrubs such as sage and woody herbs, ground covers such as sweet potato and vines such as passion fruit and kiwi fruit. The tubers consist of onions and taro.

Polyculture

Polyculture is agriculture using multiple crops in the same space, in imitation of the diversity of natural ecosystems, and avoiding large stands of single crops, or monoculture. It includes crop rotation, multi-cropping, and inter-cropping. Alley cropping is a simplification of the layered system which typically uses just two layers, with alternate rows of trees and smaller plants.

Guilds

Permaculture Guilds are groups of plants, animals and microbacteria which work particularly well together. These can be those observed in nature such as the White Oak guild which centers on the White Oak tree and includes 10 other plants. Native communities can be adapted by substitution of plants more suitable for human use. The Three Sisters of maize, squash and beans is a well known guild. The British National Vegetation Classification provides a comprehensive list of plant communities in the UK. Guilds can be thought of as an extension of companion planting.

Increase Edge

Permaculturists maintain that where vastly differing systems meet, there is an intense area of productivity and useful connections. The greatest example of this is the coast. Where the land and the sea meet there is a particularly rich area that meets a disproportionate percentage of human and animal needs. This is evidenced by the fact that the overwhelming majority of humankind lives within 100 km of the sea. So thi idea is played out in permacultural designs by using spirals in the herb garden or creating ponds that have wavy undulating shorelines rather than a simple circle or oval. Edges between woodland and open areas have been claimed to be the most productive.

Perennial Plants

Perennial plants are often used in permaculture design. As they do not need to be planted every year they require less maintenance and fertilisers. They are especially important in the outer zones and in layered systems. Ken Fern of Plants For A Future has spent many years investigating suitable perennial plants. As has Wes Jackson of The Land Institute.

Animals

Many permaculture designs involve animals other than humans. Chickens can be used as a method of weed control and also as a producer of eggs, meat and fertiliser. Some types of agroforestry systems combine trees with grazing animals.

Some projects are critical of the use of animals (see vegan organic gardening). However not all permaculture sites farm the animals. The animals are pets and can be treated as co-habitators and co-workers of the site, eating foods normally unpalatable to people such as slugs, termites, being an integral part of the pest management by eating some pests, supplying fertiliser through their droppings and controlling some weed species.

Annual Monoculture (anti-pattern)

Annual monoculture such as a wheatfield can be considered a pattern to be avoided in terms of space (height is uniform) and time (crops grow at the same rate until harvesting). During growth and especially after harvesting the system is prone to soil erosion from rain. The field requires a hefty input of fertilizers for growth and machinery for harvesting. The work is more likely to be repetitive, mechanised and rely on fossil fuels.

No pattern should be hard and fast and depending on the design considerations they can be broken. An example of this is broadscale permaculture practiced at Ragmans Lane Farm, which has a component of annual farming. Here the amount of human involvement is a key factor influencing the design.

Energy

Applying these values means using fewer non-renewable sources of energy, particularly petroleum based forms of energy. Burning fossil fuels contributes to greenhouse gases and global warming; however, using less energy is more than just combatting global warming. Food production should be a fully renewable system; but using current agricultural systems this is not the case. Industrial agriculture requires large amounts of petroleum, both to run the equipment, and to supply pesticides and fertilizers. Permaculture is in part an attempt to create a renewable system of food production that relies upon minimal amounts of energy.

For example permaculture focuses on maximizing the use of trees (agroforestry) and perennial food crops because they make a more efficient and long term use of energy than traditional seasonal crops. A farmer does not have to exert energy every year replanting them, and this frees up that energy to be used somewhere else.

Traditional pre-industrial agriculture was labor intensive, industrial agriculture is fossil fuel intensive and permaculture is design and information intensive and petrofree. Partially permaculture is an attempt to work smarter, not harder; and when possible the energy used should come from renewable sources such as wind power, passive solar designs or biofuels.

A good example of this kind of efficient design is the chicken greenhouse. By attaching the chicken coop to a greenhouse you can reduce the need to heat the greenhouse by fossil fuels, as the chicken's bodies heat the area. The chickens scratching and pecking can be put to good use to clear new land for crops. Their manure can be used to fertilise the soil. Feathers could be used in compost or as a mulch. In a conventional factory situation all these chicken outputs are seen as a waste problem. So in factories cooled by huge air conditioners, the chicken waste is extracted. All the energy is focused on egg production. Thus it is a further principle of permaculture that "pollution is energy in the wrong place".

Holmgren's 12 Design Principles

These restatements of the principles of permaculture appear in David Holmgren's *Permaculture: Principles and Pathways Beyond Sustainability*; Also see permacultureprinciples.com;

1. Observe and interact-By taking the time to engage with nature we can design solutions that suit our particular situation.
2. Catch and store energy-By developing systems that collect resources when they are abundant, we can use them in times of need.
3. Obtain a yield-Ensure that you are getting truly useful rewards as part of the work that you are doing.
4. Apply self-regulation and accept feedback-We need to discourage inappropriate activity to ensure that systems can continue to function well.
5. Use and value renewable resources and services-Make the best use of nature's abundance to reduce our consumptive behaviour and dependence on non-renewable resources.
6. Produce no waste-By valuing and making use of all the resources that are available to us, nothing goes to waste.
7. Design from patterns to details-By stepping back, we can observe patterns in nature and society. These can form the backbone of our designs, with the details filled in as we go.
8. Integrate rather than segregate-By putting the right things in the right place, relationships develop between those things and they work together to support each other.
9. Use small and slow solutions-Small and slow systems are easier to maintain than big ones, making better use of local resources and producing more sustainable outcomes.
10. Use and value diversity-Diversity reduces vulnerability to a variety of threats and takes advantage of the unique nature of the environment in which it resides.
11. Use edges and value the marginal-The interface between things is where the most interesting events take place. These are often the most valuable, diverse and productive elements in the system.

12. Creatively use and respond to change-We can have a positive impact on inevitable change by carefully observing, and then intervening at the right time.

Design for Ecologinomic (ecology-economic) Ethics

A basic principle is thus to "add value" to existing crops. A permaculture design therefore seeks to provide a wide range of solutions by including its main ethics (see above) as an integral part of the final value-added design. Crucially, it seeks to address problems that include the economic question of how to either make money from growing crops or exchange crops for labour such as in the LETS scheme. Each final design therefore should include economic considerations as well as give equal weight to maintaining ecological balance, making sure that the needs of people working on the project are met and that no one is exploited.

Community economics requires a balance between the three aspects that comprise a community: justice, environment and economics, also called the "triple bottom line", or "ecological-economics-ethics" (EEE) or "triple E". A cooperative farmer's market could be an example of this structure. The farmers are the workers and owners. Additionally, all economics are limited by their ecology. No economic system stands apart independently from its eco-system; therefore, all external costs must be considered when discussing economics.

Examples of Ecolonomic Design

One way of doing this is through designing a system that has "multiple outputs". For example, a wheat field interspersed with walnuts will reduce soil erosion, act as a windbreak and provide a walnut crop as well as a wheat crop. Managing two crops will be more interesting work. Here the system comes into conflict with conventional agriculture and economics. Interplanting trees in a wheat field reduces the wheat yield and makes the field harder to harvest using machinery, as the operator has to drive around the trees. Most farms specialise in a few crops at a time and seek to maximise surplus in order to increase profit. This surplus can only be maintained with a massive injection of fossil fuels.

Critiques

John Robin has been one the strongest critics of permaculture, criticising it for its potential to spread environmental weeds. This reflects a divide between native plant advocates and permaculture.

Permaculture in the tropics, as expressed in 'Permaculture: A Designer's Manual', did not produce significant amounts of food or fruit when applied in Northern New South Wales and Queensland, largely since the closed canopy is not conducive to fruit production. The system, whilst very healthy in and of itself, yielded very little produce.

Bill Mollison himself has criticised *itinerant teachers* of permaculture, who go on to teach after only a short course. At one point, Mollison unsuccessfully tried to trademark the term *permaculture* to prevent this practice.

Perhaps the strongest criticism of permaculture is found in the Review of Toby Hemenway's book *Gaia's Garden*, published in the Winter 2001 edition of the *Whole Earth Review*. In it, Greg Williams critiques the view that woods were more productive than farmland, based on the theory of ecological succession which says that net productivity declines as ecosystems mature.

He also criticised the lack of scientifically respectable data and questions whether permaculture is applicable to more than a small number of dedicated people. But Hemenway's response in the same magazine disputes Williams's claim on productivity as focusing on climax rather than on maturing forests, citing data from ecologist Robert Whittaker's book *Communities and Ecosystems*. Hemenway is also critical of Williams's characterisation of permaculture as simply forest gardening.

Contemporary Examples

In the years since its conception, permaculture has become a successful approach to designing sustainable systems. Its adaptability and emphasis on meeting human needs means that it can be utilized in every climatic and cultural zone. However, at the moment the large proportion of practitioners are only likely to be inspired individuals and there is a distinct lack of broadscale permaculture projects. Nevertheless, permaculture has also been used successfully as a development tool to help meet the needs of indigenous communities facing degraded standards of living from exposure to free-market economics.

Below are some examples of permaculture sites (see also a wiki-map of permaculture sites) :

Africa

Zimbabwe has 60 schools designed using permaculture, with a national team working within the schools' curriculum development unit. The UN High Commissioner for Refugees (UNHCR) has produced a report on using permaculture in refugee situations after successful use in camps in Southern Africa and Republic of Macedonia. The Biofarming approach applied in Ethiopia has very similar features and can be considered permaculture. It is mainly promoted by the non-governmental organisation BEA, based in Addis Ababa.

OCEANIA

Australia

The development of permaculture co-founder David Holmgren's home plot at Melliodora, Central Victoria, has been well documented at his website and published in e-book format.

Designed from permaculture principles, Crystal Waters is a socially and environmentally responsible, economically viable rural subdivision north of Brisbane, Australia. Crystal Waters was designed by Max Lindegger, Robert Tap, Barry Goodman and Geoff Young, and established in 1987. It received the 1996 World Habitat Award (assessed by Dr Wally N'Dow) for its "pioneering work in demonstrating new ways of low impact, sustainable living". Eighty-three freehold residential and two commercial lots occupy 20% of the 259ha (640 acre) property. The remaining 80% is the best land, and is owned in common. It can be licensed for sustainable agriculture, forestry, recreation and habitat projects.

Tikopia

Tikopians practice an intensive permaculture system, similar in principle to forest gardening and the gardens of the New Guinea highlands. Their agricultural practices are strongly and consciously tied to the population density. For example, around 1600 AD, the people agreed to slaughter all pigs on the island and substitute fishing, because the pigs were taking too much food that could be eaten by people.

New Zealand

There are many well established living examples of permaculture practice in New Zealand. Rainbow Valley Farm is the premier model. Rainbow Valley Farm was established in 1988 by Joe Polaischer and Trish Allen. The 21 ha. organic farm was designed on permaculture principles and ethics.

ASIA

Indonesia

The Indonesian Development of Education and Permaculture assisted in disaster relief in Aceh, Indonesia after the 2004 Tsunami. They have also developed Wastewater Gardens, a small-scale sewage treatment systems similar to Reedbeds.

Thailand

The Panya Project, located in Mae Taeng, Chiang Mai, Thailand, is a sustainable living project implementing permaculture principals and hosting workshops in English and Thai. In fall 2006, the project hosted a PDC taught by Geoff Lawton of the Permaculture Research Institute of Australia, and subsequently installed over 500 meters of swales and a 2 million liter dam.

The Panya Project used permaculture to help regenerate what used to be a monocrop mango plantation, transforming it into what is called a "biodiverse food forest, organic farm and education center". The Panya Project also incorporates what they call "natural building" into their design, e.g., wattle, cob and adobe brick.

EUROPE

Cyprus

Two acres of land at Ayia Skepi Therapeutic Centre in Filani village, a drug rehabilitation centre about 25 km from Nicosia, are being developed by Emily Markides, Julia Yelton, Charles Yelton and the residents of the detoxification centre.

France

- Annual events
  * Introduction to Permaculture events
  * Permaculture design courses in French and English
  * Regular 'entre-aide' work days.
- Francophone Permaculture
- Francophone Permaculture Forum
- L'Université Populaire de Permaculture
- The French Permaculture Association

2009 will see a *Permaculture Festival* with music, stories, film, accreditations, workshops, discussions and much more.

United Kingdom

There are a number of example permaculture projects in the UK, including:

- Agroforestry Research Trust, a not-for-profit organisation based in Dartington, Devon that runs a 2-acre (8,100 m^2) forest garden and publishes the journal Agroforestry News
- Chickenshack Housing co-op, a fully mutual housing co-op established in 1995 using permaculture design principles. Based in rural North Wales, the community has 4 dwellings and 6 residents on a 5-acre (20,000 m^2) site. Features include a biomass and solar district heating scheme, a half-acre forest garden and various wildlife conservation and habitat creation strategies. The community is very active in regional sustainability projects such as the Machynlleth Transition Towns initiative. It runs occasional courses in permaculture

design and regularly receives visits from interested parties.

- Middlewood Trust, a permculture-based farm in North Lancashire running courses in permaculture, crafts, forestry and sustainability
- Plants for a Future, a vegan-organic project based at Lostwithiel in Cornwall that is researching and trialing edible and otherwise useful plant crops for sustainable cultivation. Their online database features over 7,000 such species that can be grown within the UK. A collaborative version of the database is in development by the permaculture.info project.
- *Prickly Nut Woods*, a 10-acre (40,000 m^2) woodland near Haslemere, Surrey that is managed by Ben Law. He uses a 'whole system' permacultural approach, using a wide variety of woodland products and documenting a complex web of relationships. He built a house almost entirely using products from the woodland, which was featured in Channel 4's Grand Designs TV series. The project has a second, larger property in North Devon, for which it is seeking a new group to take over.
- Ragmans Lane, a 60-acre (240,000 m^2) farm in the Forest of Dean in Gloucestershire.
- The RISC Roof Garden, on top of a development education centre in Reading city centre and inspired by Robert Hart's permaculture forest garden in Shropshire, is an excellent example of urban permaculture design.. It is used by schools, educators and designers as an educational resource for sustainable development and is a member of the National Gardens Scheme. The garden is composed of dense plantings of over 180 species of edible and medicinal plants and is fed by rainwater and composted waste from the centre.
- Tir Penrhos Isaf, near Dolgellau, developed by Chris and Lyn Dixon since 1986.

Other projects tend to be more community oriented, particularly in urban areas. These include Naturewise, a north

London based group that tends a number of forest gardens and allotments as well as running regular permaculture introductory and design courses; and Organiclea, a workers cooperative that is involved in developing local food-growing and distribution initiatives around the Walthamstow area of east London. The Transition Towns movement initiated in Totnes and Kinsale by Rob Hopkins is underpinned by permaculture design principles in its attempts to visualise sustainable communities beyond peak oil.

The UK Permaculture Association publishes an extensive directory of other projects and example sites throughout the country.

NORTH AMERICA

Canada

Kootenay Permaculture Institute, British Columbia http://www3.telus.net/permaculture

United States of America

- Permaculture Research Institute Minnesota
- In Santa Fe, New Mexico, the Permaculture Institute utilizes a hands on approach to education on topics such as landscape and building design as well as water systems.
- Promoting urban permaculture in Los Angeles is Path to Freedom
- The Edible Plant Project implements and promotes elements of permaculture through a nonprofit nursery and workshops in Gainesville, FL (home of University of Florida).
- The Northeastern Permaculture Network brings together enthusiasts in the northeastern United States and eastern Canada.
- The Seattle Permaculture Guild is active in that city.
- The Portland Permaculture Guild (PPG) is very active in Portland, Oregon. There are many PC gardens in and around the Portland area. There are many fine teachers

in Portland and in Oregon, including teachers OF PC teachers. Also, Toby Hemenway, noted Permaculture author and teacher, lives in the Portland area.

- The Regenerative Design Institute (RDI) is a non-profit educational organization in Bolinas, California.
- Burlington Permaculture is an ad-hoc community group based out of Burlington Vermont. BP unites neighbors to promote urban agriculture and reforestation, enhance neighborhoods, and strengthen the web of community resources as we look beyond sustainability towards a vibrant healthy relationship with our landscape.
- The Round Mountain Institute in the Gunnison valley of Colorado is a nonprofit that is dedicated to sustainable agriculture high in the rocky mountains near the continental divide.
- The Urban Permaculture Guild implements and promotes elements of permaculture through educational workshops and projects in East Bay and San Francisco, CA.
- The Urban-Suburban Sustainability Initiative based in Belleville, Illinois is a local grassroots organization in the process of starting up. It's focus will be on permaculture, bioremedification, environmental education and Local Exchange Trading Systems.
- Indigenous Permaculture (IPP) revitalizes the relationship of communities to the earth, and operates as a collaborative of communities sharing information, resources, and tools.

Cuba

Cuba has in the past 18 years transformed its food production using low-input, or organic agriculture and, to some degree, permaculture. Havana produces up to 50% of its food requirements within the city limits, all of it is organic and produced by people in their homes, gardens and in municipal spaces.

LATIN AMERICA

Nicaragua

Project Bona Fide is a 43-acre (170,000 m^2) site on the twin volcano island of Ometepe, Nicaragua. Project Bona Fide has been in development for nearly a decade, and has become an important center for education and community development. Infrastructural systems contain: natural buildings built with local materials, terraced and medicinal plant gardens, an extensive nursery, seed bank, developing fruit and nut orchards, food forests, native timber forestry, timber bamboo plantings, water-catchment, drip irrigation and ferrocement technologies, renewable energy systems, and composting toilets. Outreach efforts include social programs that provide educational opportunities based in ecological agriculture, community reforestation efforts that are supported by our seed bank and nursery, local seed and plant exchanges, a children's nutritional kitchen and an upcoming community center.

Brazil

IPEC-Ecocentro at the Instituto de Permacultura e Ecovilas do Cerrado-the Institute of Permaculture and Ecovillage of the Cerrado

IPCP-Instituto de Permacultura Cerrado-Pantanal (Permaculture Institute of Cerrado-Pantanal), Campo Grande, MS. Specializing in interactive teaching of Permaculture and direct work with Indigenous communities within the Cerrado biome.

5

Urban Agriculture

Urban agriculture is the practice of cultivating, processing and distributing food in, or around (peri-urban), a village, town or city.

Urban farming is generally practiced for income-earning or food-producing activities though in some communities the main impetus is recreation and relaxation. Urban agriculture contributes to food security and food safety in two ways: first, it increases the amount of food available to people living in cities, and, second, it allows fresh vegetables and fruits and meat products to be made available to urban consumers. A common and efficient form of urban agriculture is the biointensive method. Because urban agriculture promotes energy-saving local food production, urban and peri-urban agriculture are generally seen as sustainable practices.

The recognition of environmental degradation within cities through the relocation of resources to serve urban populations has inspired the implementation of different schemes of urban agriculture across the developed and developing world. From historic models such as Machu Picchu to designs for new productive urban farms, the idea of locating agriculture in the city takes on many characteristics.

History of Urban Agriculture

Community wastes were used in ancient Persia to feed urban farming. In Machu Picchu water was conserved and reused as part of the stepped architecture of the city and vegetable beds were designed to gather sun in order to prolong the growing season. Victory gardens sprouted during WWI, WWII and were fruit, vegetable, and herb gardens in US, Canada, and UK. This effort was undertaken by citizens to reduce pressure on food production that was to support the war effort. Community gardening in most communities are open to the public and provide space for citizens to cultivate plants for food or recreation. A community gardening program that is well-established is Seattle's P-Patch.

Urban Agricultural Facts

- 50% of the world's population lives in cities.
- 800 million people are involved in urban agriculture world-wide and contribute to feeding urban residents.
- Low income urban dwellers spend between 40% and 60% of their income on food each year.
- By 2015 about 26 cities in the world are expected to have a population of 10 million or more. To feed a city of this size – at least 6000 tonnes of food must be imported each day.

Perspectives on Urban Agriculture

Resource and Economic

The Food and Agriculture Organization of the United Nations (FAO), has defined urban agriculture as:

" [A]n industry that produces, processes and markets food and fuel, largely in response to the daily demand of consumers within a town, city, or metropolis, on land and water dispersed throughout the urban and peri-urban area, applying intensive production methods, using and reusing natural resources and urban wastes to yield a diversity of crops and livestock."

The definition of urban agriculture as an industry that responds to the nutritional demands of a city, from within that city, with the use and reuse of that city's resources while acknowledging economic and resource use does not reconcile aspects of regional health, food security, and application of grassroots organizations. (This definition is based on the work of Luc Mougeot of the International Development Research

Centre and used in technical and training publications by UN-HABITAT's Urban Management Programme, FAO's Special Programme for Food Security, and international agricultural research centres, such as CIRAD.)

Environmental

The Council on Agriculture, Science and Technology, (CAST) is an international consortium of scientific and professional societies based in Ames Iowa that compiles and communicates credible science-based information to policy makers, media, private sector, and the public. CAST defines urban agriculture to include aspects of environmental health, remediation, and recreation,

"Urban agriculture is a complex system encompassing a spectrum of interests, from a traditional core of activities associated with the production, processing, marketing, distribution, and consumption, to a multiplicity of other benefits and services that are less widely acknowledged and documented. These include recreation and leisure; economic vitality and business entrepreneurship, individual health and well-being; community health and well being; landscape beautification; and environmental restoration and remediation."

Modern planning and design initiatives are more responsive to this model of urban agriculture because it fits within the current scope of sustainable design. The definition allows for a multitude of interpretations across cultures and time. Frequently it is tied to policy decisions to build sustainable cities.

Food Security

Access to nutritious food is another perspective in the effort to locate food and livestock production in cities. With the tremendous influx of world population to urban areas, the need for fresh and safe food is increased. Community Food Security Coalition (CFSC) defines food security as, "all persons in a community having access to culturally acceptable, nutritionally adequate food through local, non-emergency sources at all times."

The Emerging Importance of Urban Agriculture

Economic

- UPA (urban and peri-urban agriculture) expands the economic base of the city through production, processing, packaging, and marketing of consumable products. This results in an increase in entrepreneurial activities and the creation of job opportunities, as well as in food costs reduction and products of better quality.
- UPA represents an important opportunity for women to be part of the informal economy of a city. Farming and selling activities can be combined more easily with household tasks and child care.
- UPA provides employment, income, and access to food for urban populations, which together contributes to relieve chronic and emergency food insecurity. Chronic food insecurity refers to less affordable food and growing urban poverty, while emergency food insecurity relates to breakdowns in the chain of food distribution. UPA plays an important role in making food more affordable and in providing emergency supplies of food. Research into market values for produce grown in urban gardens has attributed to a community garden plot a median yield value of between approximately $200 and $500 (US, adjusted for inflation). In a community gardening program as well-established as Seattle's P-Patches, this can account for up to 1.25 million dollars of produce cultivated annually.

Social

Social benefits that have emerged from urban agricultural practices are; better health and nutrition, increased income, employment, food security within the household, and community social life. UPA can be seen as a means of improving the livelihood of people living in and around cities. Taking part in such practices is seen mostly as informal activity, but in many cities where inadequate, unreliable, and irregular access to food is an occurring problem, urban agriculture has been a positive response to tackling food concerns. Households and small communities take advantage of vacant land and contribute not only to their household food needs but also the needs of their resident city. The CFSC states that, "Community and residential gardening, as well as small-scale farming, save household food dollars. They promote nutrition and free cash for non-garden foods and other items"

This allows families to generate larger incomes selling to local grocers or to local outdoor markets, while supplying their household with proper nutrition of fresh and nutritional produce.

Some community urban farms can be quite efficient and help women find work, who in some cases are marginalized from finding employment in the formal economy. Studies have shown that participation from women have a higher production rate, therefore producing the adequate amount for household consumption while supplying more for market sale.

Due to the fact that most UA activities are conducted on vacant municipal land, there have been rising concerns about the allocation of land and property rights. The IDRC and the FAO have published the Guidelines for Municipal Policymaking on Urban Agriculture, and are working with municipal governments to create successful policy measures that can be incorporated in urban planning. Including UA in local plans and as proper land use will continue to help impoverished communities gain a better well-being while fighting urban poverty. Localized food production in urban and peri-urban areas contributes to local economies by creating jobs and producing valuable products. Some researchers indicate that unemployed populations in large cities and suburban towns would decrease if put to work by local food movements. Schools have foreseen the asset of local food production and are beginning to incorporate agricultural sections in their curricula and present it as a career opportunity. Urban agricultural projects are beginning to open a new labor market in areas that have been negatively affected by industrial outsourcing of jobs.

Energy Efficiency

The current industrial agriculture system is accountable for high energy costs for the transportation of foodstuffs. The average conventional produce item travels 1,500 miles, using, if shipped by tractor-trailer, one gallon of fossil fuel per hundred pounds. The energy used to transport food is decreased when urban agriculture can provide cities with locally-grown food.

Quality of Food

Although the taste of locally grown food is subjective, many participants in the urban agriculture movement report that they prefer the taste of local agricultural products, or organic food, to that of industrial food production. Also, urban agriculture supports a more sustainable production of the food that tries to decrease the use of harmful pesticides that result in agricultural runoff. Urban and local farmers also eliminate the need for preservatives, as their products do not need to travel long distances.

Economy of Scale

Using high-density urban farming, as for instance with vertical farms or stacked greenhouses, many environmental benefits can be achieved on a city-wide scale that would be impossible otherwise. These systems do not only provide food, but also produce potable water from waste water, and can recycle organic waste back to energy and nutrients. At the same time, they can reduce food-related transportation to a minimum while providing fresh food for large communities in almost any climate.

IMPLEMENTATION OF URBAN AGRICULTURE

Community-based Infrastructure

Creating a community-based infrastructure for urban agriculture means establishing local systems to grow and process food and transfer it from farmer (producer) to consumer.

To facilitate food production, cities have established community-based farming projects. Some projects have collectively-tended community farms on common land, much like that of eighteenth-century Boston Common. One such community farm is the Collingwood Children's Farm in Melbourne, Australia. Other community garden projects use the allotment garden model, in which gardeners care for individual plots in a larger gardening area, often sharing a tool shed and other amenities. Seattle's P-Patch gardens use this model, as did the South Central Farm in Los Angeles. Independent urban gardeners also grow food in individual yards and on roofs. Roof gardens allow for urban dwellers to maintain green spaces in the city without having to set aside a tract of undeveloped land. There are a growing number of projects worldwide that seek to enable cities to become 'continuous productive landscapes' through the networked cultivation of vacant urban land and temporary or permanent 'kitchen gardens'. Food processing on a community level has been accommodated by centralizing resources in community tool sheds and processing facilities for farmers to share. The Garden Resource Program Collaborative based in Detroit has cluster tool banks. Different areas of the city have toolbanks where resources like tools, compost, mulch, tomato stakes, seeds, and education can be shared and distributed with the gardeners in that cluster. Detroit's Garden Resource Program Collaborative also strengthens their gardening community by providing to their members transplants; education on gardening, policy, and food issues; and by building connectivity between gardeners through workgroups, potlucks, tours, field trips, and cluster workdays.

Farmers' markets, such as the farmers' market in Los Angeles, provide a common land where farmers can sell their product to consumers. Large cities tend to open their farmers markets on the weekends and one day in the middle of the week. For example, the farmers' market of Rue Richard Lenoir in Paris, France, is open on Sundays and Thursdays. However, to create a consumer dependency on urban agriculture and to introduce local food production as a sustainable career for farmers, markets would have to be open regularly. For example, the Los Angeles Farmers' Market is open seven days a week and has linked several local grocers together to provide different food products. The market's central location in downtown Los Angeles provides the perfect interaction for a diverse group of sellers to access their consumers.

Havana, Cuba

Due to the shortage of fuel and therefore severe deficiencies in the transportation sector a growing percentage of the agricultural production takes place in the so-called urban agriculture. In 2002, 35,000 acres (140 km^2) of urban gardens produced 3.4 million tons of food. In Havana, 90% of the city's fresh produce come from local urban farms and gardens. In 2003, more than 200,000 Cubans worked in the expanding urban agriculture sector.

Mumbai, India

Economic development in Mumbai brought a growth in population caused mainly by the migration of laborers from other regions of the country. The number of residents in the city increased more than twelve times in the last century. Greater Mumbai, formed by City Island and Salsette Island, is the largest city in India with a population of 16.4 million, according to data collected by the census of 2001. Mumbai is one of the densest cities in the world, 48,215 persons per km^2 and 16,082 per km^2 in suburban areas. In a scenario like this, urban agriculture seems unlikely to put in practice since it must compete with real estate developers for the access and use of vacant lots. Alternative farming methods have emerged as a response to scarcity of land, water, and economic resources employed in UPA.

Dr. Doshi's innovative techniques on farming: Dr. Doshi's city garden methods are revolutionary for being appropriate to apply in reduced spaces as terraces and balconies, even on civil construction walls, and for not requiring big investments in capital or long hours of work. His farming practice is purely organic and is mainly directed to domestic consumption. His gardening tools are composed of materials available in the local environment: sugarcane waste, polyethylene bags, tires, containers and cylinders, and soil. The containers and bags (open at both ends) are filled with the sugarcane stalks, compost, and garden soil, which make possible the use of minimal quantity of water if compared to open fields. Dr. Doshi states that solar energy can replace soil in cities. He also recommends the idea of chainplanting, or growing plants in intervals and in small quantities rather than at once and in large amounts. He has grown different types of fruit such as mangos, figs, guavas, bananas, and sugarcane stalks in his terrace of 1,200 sq ft (110 m^2) in Bandra. The concept of city farming developed by Dr. Doshi consumes the entire household's organic waste. He subsequently makes the household self-sufficient in the provision of food: 5 kg of fruits and vegetables are produced daily for 300 days a year.

City farm at Rosary High School, Dockyard Road Mumbai: The main objectives of this pilot project were to promote economic support for street children, beautify the city landscape, supply locally produced organic food to urban dwellers (mainly those residing in slums), and to manage organic waste in a sustainable city. The project was conducted in the Rosary School, in Mumbai, with the participation of street children during 2004. A city farm was created in a terrace area of 400 sq ft (37 m^2). The participants were trained in urban farming techniques. The farm produced vegetables, fruits, and flowers. The idea was spread the concept of city farm to other schools in the city.

Mumbai Port Trust (MBPT): The central kitchen distributes food to approximately 3,000 employees daily, generating important amounts of organic disposal. A terrace garden created by the staff recycles ninety percent of this waste in the production of vegetables and fruits. Preeti Patil, who is the catering officer at the MBPT explains the purpose of the enterprise:

"Mumbai Port Trust has developed an organic farm on the terrace of its central kitchen, which is an area of approximately 3,000 sq ft (280 m^2). The activity of city farming was started initially to dispose of kitchen organic waste in an ecofriendly way. Staff members, after their daily work in the kitchen, tend the garden, which has about 450 plants."

Bangkok, Thailand

In early 2000, urban gardens were started under the direction of the NGO, Thailand Environment Institute (TEI), to help achieve the Bangkok Metropolitan Administrations

(BMA) priority to 'green' Thailand. With a population of 12 million and 39% of the land in the city vacant due to rapid expansion of the 1960's-80's Bangkok is a test bed for urban gardens centered on community involvement. The two urban gardens initiated by TEI are in Bangkok Noi and Bangkapi and the main tasks were stated as:

- Teach members of the communities the benefits of urban green space.
- Create the social framework to plan, implement, and maintain the urban green space.
- Create a process of method to balance the needs of the community with the needs of the larger environmental concerns.

While the goals of the NGO are important in a global context, the community goals are being met through the work of forming the urban gardens themselves. In this sense the creation, implementation, and maintenance of urban gardens is highly determined by the desires of the communities involved. However, the criteria by which TEI measured their success illustrates the scope of benefits to a community which practices urban agriculture. TEI's success indicators were:

- Establishing an Urban Green Plan
- Community Capacity Building
- Poverty Reduction
- Links with Government
- Status of Women<
- Developing a Model for Other Communities

Evan D.J. Fraser wrote in the article Urban Ecology in Bangkok Thailand that although the project was initiated to serve the environmental needs of the city it quickly illustrated the positive side effects of urban agriculture.

"In many ways, the urban environment became a lens through which communities re-evaluated their own relationship with the city, the impact of urbanization in a global context, and how small groups can exert some control over the shape of their neighbourhoods."

Beijing, China

Beijing's increase in land area from 4,822 km^2 in 1956 to 16,808 km^2 in 1958 led to the increased adoption of peri-urban agriculture. Such "suburban agriculture" led to more than 70% of non-staple food in Beijing, mainly consisting of vegetables and milk, to be produced by the city itself in the 1960s and 1970s. Recently, with relative food security in China, periurban agriculture has led to improvements in the quality of the food available, as opposed to quantity. One of the more recent experiments in urban agriculture is the Modern Agricultural Science Demonstration Park in Xiaotangshan.

Shenzhen, China

Traditionally Chinese cities have been known to mix agricultural activities within the urban setting. Shenzhen, what was once a small farming community is now a fast growing metropolis due to the Chinese government designation as an open economic zone. Due to large and growing population in China, the government supports urban self-sufficiency in food production. Shenzhen's village structure, sustainable methods, and new agricultural advancements initiated by the government have been strategically configured to supply food for this growing city.

Village Structure, Methods, and New Advancements

- The city farms are located about 10 km from city center in a two-tier system. The first tier approached from city center produces perishable items. Located just outside these farms, hardier vegetables are grown such as potatoes, carrots, and onions. This system allows produce to be sold in city markets just a few short hours after picking.
- Another impressive method used within Chinese agriculture and aquaculture practice is the mulberry-dyke fish-pond system, which is a response to waste recycling and soil fertility. This system can be described as, "mulberry trees are grown to feed silkworms and the silkworm waste is fed to the fish in ponds. The fish also feed on waste from other animals, such as pigs, poultry,

and buffalo. The animals in turn are given crops that have been fertilized by mud from the ponds. This is a sophisticated system as a continuous cycle of water, waste and food...with man built into the picture"

- As population grows and industry advances the city tries to incorporate potential agricultural growth by experimenting in new agricultural methods. The Fong Lau Chee Experimental Farm in Dongguan, Guangdong has worked with new agricultural advancements in lychee production. This farm was established with aspirations of producing large quantities and high quality lychees, by constantly monitoring sugar content, and their seeds. This research, conducted by local agricultural universities allows for new methods to be used with hopes of reaching the needs of city consumers.

Government Involvement

- However due to increased levels of economic growth and pollution some urban farms have become threatened. The government has been trying to step in and create new technological advancements within the agricultural field to sustain levels of urban agriculture.
- "The city plans to invest 8.82 billion yuan in 39 agricultural projects, including a safe agricultural base, an agricultural high-tech park, agricultural processing and distribution, forestry, eco-agricultural tourism, which will form an urban agriculture with typical Shenzhen characteristics" in conjunction with this program the city is expected to expand the Buji Farm Produce Wholesale Market.
- According to the Municipal Bureau of Agriculture, Forestry and Fishery the city will invest 600 million yuan on farms located around the city, with hopes of the farms to provide "60 percent of the meat, vegetables and aquatic products in the Shenzhen market".
- There has also been an emerging trend of going green and organic as a response to pollution and pesticides used in farming practices. Vegetable suppliers are

required to pass certaın inspections held by the city's Agriculture Bureau before they can be sold as "green".

New York, NY

In New York City many low-income residents suffer from high rates of obesity and diabetes, limited sources of fresh produce and available, undeveloped land. The City and local nonprofit groups have been providing land, training and financial encouragement, but the impetus has in urban farming has really come from the farmers, who often volunteer when their regular work day is done. Some urban gardeners have used empty lots to start community or urban garden. However, the soil must be tested for heavy contamination in city soil because of vehicle exhaust and remnants of old construction. However, studies have found that such ground can be cultivated as long as the pH is kept neutral. The City also has a composting program, which is available to gardeners and farmers. One group, GreenThumb, provides free seedlings. Another program, the City Farms project operated by the nonprofit Just Food, profits courses on growing and selling food.

Benefits of Urban Agriculture

The benefits that UPA brings along to cities that implement this practice are numerous. The transformation of cities from only consumers of food to generators of agricultural products contributes to sustainability, improved health, and poverty alleviation.

- UPA assists to close the open loop system in urban areas characterized by the importation of food from rural zones and the exportation of waste to regions outside the city or town.
- Wastewater and organic solid waste can be transformed into resources for growing agriculture products: the former can be used for irrigation, the latter as fertilizer.
- Vacant urban areas can be used for agriculture production.
- Other natural resources can be conserved. The use of wastewater for irrigation increases the availability of freshwater for drinking and household consumption.

- UPA can help to preserve bioregional ecologies from being transformed into cropland.
- Urban agriculture saves energy (e.g. energy consumed in transporting food from rural to urban areas).
- Local production of food also allows savings in transportation costs, storage, and in product loss, what results in food cost reduction.
- UPA improves the quality of the urban environment through greening and thus, a reduction in pollution.

Health Improvement

- Urban agriculture also makes of the city a healthier place to live by improving the quality of the environment.
- UPA is a very efficient tool to fight against hunger and malnutrition since it facilitates the access to food by an impoverished sector of the urban population.

Poverty Alleviation: It is known that a large part of the people involved in urban agriculture is the urban poor. In developing countries, the majority of urban agricultural production is for self-consumption, with surpluses being sold in the market. According to the FAO (Food and Agriculture Organization of the United Nations), urban poor consumers spend between 60 and 80 percent of their income on food, making them very vulnerable to higher food prices.

- UPA provides food and creates savings in household expenditure on consumables, thus increasing the amount of income allocated to other uses.
- UPA surpluses can be sold in local markets, generating more income for the urban poor.

Community centers and gardens educate the community to see agriculture as an integral part of urban life. The Florida House Institute for Sustainable Development in Sarasota, Florida, serves as a public community and education center in which innovators with sustainable, energy-saving ideas can implement and test them. Community centers like Florida House provide urban areas with a central location to learn about urban agriculture and to begin to integrate agriculture with the urban lifestyle. Other examples of community centers are Greensgrow Farm in Philadelphia and Northey Street City Farm in Brisbane, Australia. Greensgrow uses an abandoned site as an urban farm to teach the community how food is grown and how to grow their own food. Northey Street City Farm hosts weekly community activities to educate and involve local residents in agricultural practices. Urban farms also are a proven effective educational tool to teach kids about healthy eating and meaningful physical activity. An example of educational urban agriculture is Full Circle Farm, an 11-acre farm located on a middle school campus in the heart of Silicon Valley. Like other educational agriculture centers, Full Circle Farm's acreage is used as a "living campus" where students get real-world, hands-on agriculture experiences that cultivate both healthy habits and environmental leadership.

Difficulties of Urban Agriculture

- The use of waste water for irrigation without careful treatment and monitoring can result in the spread of diseases among the population.
- Cultivation on contaminated land represents also a health hazard for the consumers.
- The practice of cultivating along roadsides facilitates the distribution of products to local markets, but it is at the same time a risky practice since it exposes food to car pollution.
- Agriculture and urbanization are considered to be incompatible activities, competing for the access and use of limited land. In reality, in urban areas there is important available space for agriculture use such as public and private vacant lots, and areas not suited for built-up uses (steep slopes and flood plains).
- Legal restrictions and economic impediments to accessing land and resources (such as reasonably priced water) are among the most common problems confronted by urban agriculture.
- Lack of security of tenure also acts as a preventive for farming due to the uncertainty in the use length of the land.

- Urban agriculture has been criticized by those who believe that industrial farm production can produce food at larger volumes more efficiently.
- A major argument is whether urban farming alone-that is farming very intensively on small land areas-could replace land extensive production in rural areas which produce the bulk of our food products. Yet, hunger persists in both urban and rural areas (see more on food security), despite a subsidized industrial agriculture. The degree to which urban agriculture can address these food needs systemically is undetermined, though there are indications in some communities it is an important source of food.
- Other opponents argue that localized food production and the introduction of common resources and common lands into the urban areas would produce a tragedy of the commons. Though, as referenced earlier, many urban farms and community gardens are managed privately or through other civil society organizations.

NEOLITHIC REVOLUTION

The Neolithic Revolution was the first agricultural revolution—the transition from hunting and gathering communities and bands, to agriculture and settlement (settlement is currently being questioned). Archaeological data indicate that various forms of domestication of plants and animals arose independently in at least 7-8 separate locales worldwide, with the earliest known developments taking place in the Middle East around 10,000 BC (BCE) or earlier.

However, the Neolithic Revolution involved far more than the adoption of a limited set of food-producing techniques. During the next millennia it would transform the small, mobile and fairly egalitarian groups of hunter-gatherers that had hitherto dominated human history, into sedentary societies based in built-up villages and towns, which radically modified their natural environment by means of specialized cultivation and storage technologies (e.g. irrigation) that allowed extensive surplus production. These developments provided the basis for high population densities, complex labor diversification, trading economies, centralized administrations and political structures, hiearchical ideologies and depersonalized systems of knowledge (e.g. property regimes and writing). The first full-blown manifestation of the entire Neolithic complex is seen in the Middle Eastern Sumerian cities (ca. 5,300 BC), whose emergence also inaugurates the end of the prehistoric Neolithic and the beginning of historical time.

The relationship of the above-mentioned Neolithic characteristics to the onset of agriculture, their sequence of emergence and empirical relation to each other at various Neolithic sites remains the subject of academic debate, and seems to vary from place to place, rather than being the outcome of universal laws of social evolution.

Agricultural Transition

The term *Neolithic Revolution* was coined in the 1920s by Vere Gordon Childe to describe the first in a series of agricultural revolutions in Middle Eastern history. The period is described as a "revolution" to denote its importance, and the great significance and degree of change affecting the communities in which new agricultural practices were gradually adopted and refined.

The beginning of this process in different regions has been dated from perhaps 10,000 years ago in Melanesia to 2,500 BC in Subsaharan Africa, with some considering the developments of 9000-7000 BC in the Fertile Crescent to be the most important. This transition everywhere seems associated with a change from a largely nomadic hunter-gatherer way of life to a more settled, agrarian-based one, with the inception of the domestication of various plant and animal species-depending upon which species were locally available, and probably also influenced by local culture.

There are several competing (but not mutually exclusive) theories as to the factors which drove populations to take up agriculture. The most prominent of these are:

- The Oasis Theory, originally proposed by Raphael Pumpelly in 1908, popularized by Vere Gordon Childe in 1928 and summarised in Childe's book *Man Makes Himself*., which maintains that as the climate got drier, communities contracted to oases where they were forced into close association with animals, which were then domesticated together with planting of seeds. It has little support now because climate data for the time does not support the theory.

- The Hilly Flanks hypothesis, proposed by Robert Braidwood in 1948, suggests that agriculture began in the hilly flanks of the Taurus and Zagros mountains, where the climate was not drier as Childe had believed, and fertile land supported a variety of plants and animals amenable to domestication.
- The Feasting model by Brian Hayden suggests that agriculture was driven by ostentatious displays of power, such as giving feasts, to exert dominance. This required assembling large quantities of food which drove agricultural technology.
- The Demographic theories proposed by Carl Sauer and adapted by Lewis Binford and Kent Flannery posit an increasingly sedentary population which expanded up to the carrying capacity of the local environment and required more food than could be gathered. Various social and economic factors helped drive the need for food.
- The evolutionary/intentionality theory, developed by David Rindos and others, views agriculture as an evolutionary adaptation of plants and humans. Starting with domestication by protection of wild plants, it led to specialization of location and then full-fledged domestication.

In contrast to the Paleolithic, in which more than one hominid species existed, only one (Homo sapiens) reached the Neolithic.

Domestication of Plants

Once agriculture started gaining momentum, cereal grasses (beginning with emmer, einkorn and barley), and not simply those that would favour greater caloric returns through larger seeds, were selectively bred. Plants that possessed traits such as small seeds or bitter taste would have been seen as undesirable. Plants that rapidly shed their seeds on maturity tended not to be gathered at harvest, thus not stored and not seeded the following season; years of harvesting selected for strains that retained their edible seeds longer. Several plant species, the "pioneer crops" or Neolithic founder crops, were the earliest plants successfully manipulated by humans. Some of these pioneering attempts failed at first and crops were abandoned, sometimes to be taken up again and successfully domesticated thousands of years later: rye, tried and abandoned in Neolithic Anatolia, made its way to Europe as weed seeds and was successfully domesticated in Europe, thousands of years after the earliest agriculture. Wild lentils present a different challenge that needed to be overcome: most of the wild seeds do not germinate in the first year; the first evidence of lentil domestication, breaking dormancy in their first year, was found in the early Neolithic at Jerf el-Ahmar, (in modern Syria), and quickly spread south to the Netiv HaGdud site in the Jordan Valley. This process of domestication allowed the founder crops to adapt and eventually become larger, more easily harvested, more dependable in storage and more useful to the human population.

Figs, barley and, most likely, oats were cultivated in the Jordan Valley, represented by the early Neolithic site of Gilgal, where in 2006 archaeologists found caches of seeds of each in quantities too large to be accounted for even by intensive gathering, at strata dateable c. 11,000 years ago. Some of the plants tried and then abandoned during the Neolithic period in the Ancient Near East, at sites like Gilgal, were later successfully domesticated in other parts of the world.

Once early farmers perfected their agricultural techniques, their crops would yield surpluses which needed storage. Most hunter gatherers could not easily store food for long due to their migratory lifestyle, whereas those with a sedentary dwelling could store their surplus grain. Eventually granaries were developed that allowed villages to store their seeds for longer periods of time. So with more food, the population expanded and communities developed specialized workers and more advanced tools.

The process was not as linear as was once thought, but a more complicated effort, which was undertaken by different human populations in different regions in many different ways.

Agriculture in Asia

The Neolithic Revolution is believed to have become widespread in southwest Asia around 8000 BC–7000 BC, though earlier individual sites have been identified. Although archaeological evidence provides scant evidence as to which of the genders performed what task in Neolithic cultures, by comparison with historical and contemporary hunter-gatherer communities it is generally supposed that hunting was typically performed by the men, whereas women had a more significant role in the gathering. By extension, it may be theorised that women were largely responsible for the observations and initial activities which began the Neolithic Revolution, insofar as the gradual selection and refinement of edible plant species was concerned.

The precise nature of these initial observations and (later) purposeful activities which would give rise to the changes in subsistence methods brought about by the Neolithic Revolution are not known; specific evidence is lacking. However, several reasonable speculations have been put forward; for example, it might be expected that the common practice of discarding food refuse in middens would result in the regrowth of plants from the discarded seeds in the (fertilizer-enriched) soils. In all likelihood, there were a number of factors which contributed to the early onset of agriculture in Neolithic human societies.

Agriculture in the Fertile Crescent

Generalised agriculture apparently first arose in the Fertile Crescent because of many factors. The Mediterranean climate has a long dry season with a short period of rain, which made it suitable for small plants with large seeds, like wheat and barley. These were the most suitable for domestication because of the ease of harvest and storage and the wide availability. In addition, the domesticated plants had especially high protein content. The Fertile Crescent had a large area of varied geographical settings and altitudes. The variety given made agriculture more profitable for former hunter-gatherers. Other areas with a similar climate were less suitable for agriculture because of the lack of geographic variation within the region and the lack of availability of plants for domestication.

Agriculture in Africa

The Revolution developed independently in different parts of the world, not just in the Fertile Crescent. On the African continent, three areas have been identified as independently developing agriculture: the Ethiopian highlands, the Sahel and West Africa.

The most famous crop domesticated in the Ethiopian highlands is coffee. In addition, Khat, Ensete, Noog, teff and finger millet were also domesticated in the Ethiopian highlands. Crops domesticated in the Sahel region include sorghum and pearl millet. The Kola nut, extracts from which became an ingredient in Coca Cola, was first domesticated in West Africa. Other crops domesticated in West Africa include African rice, African yams and the oil palm.

A number of crops that have been cultivated in Africa for millennia came after their domestication elsewhere. Agriculture in the Nile River Valley developed from crops domesticated in the Fertile Crescent. Bananas and plantains which were first domesticated in Southeast Asia, most likely Papua New Guinea, were re-domesticated in Africa possibly as early as 5,000 years ago. Asian yams and taro were also cultivated in Africa.

Prof. Fred Wendorf and Dr. Romuald Schild, of the Department of Anthropology at Southern Methodist University, originally thought to have found evidence of early agriculture in Upper Paleolithic times at Wadi Kubbaniya, on the Kom Ombos plateau, of Egypt, including a mortar and pestle, grinding stones, several harvesting implements and charred wheat and barley grains — which may have been introduced from outside the region. AMS dating since their first reports has invalidated their hypothesis. Many such grinding stones are found with the early Egyptian Sebilian and Mechian cultures and evidence has been found of a neolithic domesticated crop-based economy dating around 5000 BC. Smith writes: "With the benefit of hindsight we can now see that many Late Paleolithic peoples in the Old World were poised on the brink of plant cultivation and animal husbandry as an alternative to the hunter-gatherer's way of life". Unlike the Middle East, this evidence appears as a "false dawn" to agriculture, as the sites were later abandoned, and permanent farming then was delayed until 4,500 BC with the Tasian and Badarian cultures and the arrival of crops and animals from the Near East.

Agriculture in the Americas

Corn, beans and squash were domesticated in Mesoamerica around 3500 BCE. Potatoes and manioc were domesticated in South America. In what is now the eastern United States, Native Americans domesticated sunflower, sumpweed and goosefoot around 2500 BCE.

Domestication of Animals

When hunter-gathering began to be replaced by sedentary food production it became more profitable to keep animals close at hand. Therefore, it became necessary to bring animals permanently to their settlements, although in many cases there was a distinction between relatively sedentary farmers and nomadic herders. The animals' size, temperament, diet, mating patterns, and life span were factors in the desire and success in domesticating animals. Animals that provided milk, such as cows and goats, offered a source of protein that was renewable and therefore quite valuable. The animal's ability as a worker (for example ploughing or towing), as well as a food source, also had to be taken into account. Besides being a direct source of food, certain animals could provide leather, wool, hides, and fertilizer. Some of the earliest domesticated animals included dogs (about 15,000 years ago), sheep, goats, cows, and pigs.

Domestication of Animals in the Middle East

The Middle East served as the source for many animals that could be domesticated, such as goats and pigs. This area was also the first region to domesticate the Dromedary Camel. The presence of these animals gave the region a large advantage in cultural and economic development. As the climate in the Middle East changed, and became drier, many of the farmers were forced to leave, taking their domesticated animals with them. It was this massive emigration from the Middle East that would later help distribute these animals to the rest of Afroeurasia. This emigration was mainly on an east-west axis of similar climates, as crops usually have a narrow optimal climatic range outside of which they cannot grow for reasons of light or rain changes. For instance, wheat does not normally grow in tropical climates, just like tropical crops such as bananas do not grow in colder climates. Some authors like Jared Diamond postulated that this East-West axis is the main reason why plant and animal domestication spread so quickly from the Fertile Crescent to the rest of Eurasia and North Africa, while it did not reach through the North-South axis of Africa to reach the Mediterranean climates of South Africa, where temperate crops were successfully imported by ships in the last 500 years. The African Zebu is a separate breed of cattle that was better suited to the hotter climates of central Africa than the fertile-crescent domesticated bovines. North and South America were similarly separated by the narrow tropical Isthmus of Panama, that prevented the andes llama to be exported to the Mexican plateau.

Causes of the Neolithic Revolution

Jack Harlan, examining the causes for the Neolithic Revolution, suggests 6 principal reasons which can be summarized to 3 principal categories:

1. Domestication for religious reasons
2. Domestication by crowding and as a consequence of stress
3. Domestication resulting from discovery, based upon the perceptions of food gatherers

With regard to the first explanation, Ian Hodder, who directs the excavations at Çatalhöyük, has said that the earliest settled communities, and the Neolithic revolution they represent, actually *preceded* the development of agriculture. He has been developing the ideas first expressed by Jacques Cauvin, the excavator of the Natufian settlement at Mureybet in northern Syria. Hodder believes that the Neolithic revolution was the result of a revolutionary change in the human psychology, a "revolution of symbols" which led to new beliefs about the world and shared community rituals embodied in corpulent female figurines (see Venus of Willendorf) and the methodical assembly of aurochs horns. An alternative explanation for the origin of agriculture has been advanced by anthropologist Mark Nathan Cohen. Cohen believes that following the widespread extinctions of large mammals in the late Palaeolithic, the human population had expanded to the limits of the available territory and a population explosion led to a food crisis. Agriculture was the only way in which it was possible to support the increasing population on the available area of land. This view has come under criticism due to the obvious problem of how a population explosion would occur without already having a surplus of food.

Food gatherers (not the hunters) caring for children, keeping the fires alive, and foraging near the base camp, led the way in developing language and culture, in knowledge of plants, and increasingly semi-domesticated animals who travelled with the nomads from camp to camp.

CONSEQUENCES OF THE NEOLITHIC REVOLUTION

Social Change

It is often argued that agriculture gave humans more control over their food supply, but this has been disputed by the finding that nutritional standards of Neolithic populations were generally inferior to that of hunter gatherers, and life expectancy may in fact have been shorter, in part due to diseases. Actually, by reducing the necessity for the carrying of children, Neolithic societies had a major impact upon the spacing of children (carrying more than one child at a time is impossible for hunter-gatherers, which leads to children being spaced four or more years apart). This increase in the birth rate was required to offset increases in death rates and required settled occupation of territory and encouraged larger social groups. These sedentary groups were able to reproduce at a faster rate due to the possibilities of sharing the raising of children in such societies. The children accounted for a denser population, and encouraged the introduction of specialization by providing diverse forms of new labor. The development of larger societies seemed to have led to the development of different means of decision making and to governmental organization. Food surpluses made possible the development of a social elite who were not otherwise engaged in agriculture, industry or commerce, but dominated their communities by other means and monopolized decision-making.

The Neolithic Revolution had many results socially and economically. Socially people could have more children and social classes emerged. Economically a surplus of food meant that people could also specialize in different things. The Neolithic Revolution was clearly the start of our society as we know it today. About twelve thousand to six thousand years ago humans started to notice changes of season and when and how to grow food. The slow process of learning how to use the land brought about many changes in the world socially. The ability to know where food was, how to grow it and when you could eat it gave humans enough security to start settling and replace their nomadic hunter gatherer ways with agriculture and villages. As humans grew better and better at farming, there became a surplus of food. This resulted in an explosion of population as demonstrated in this chart:

Year	Human Population
3000 B.C.	14 million
2000 B.C.	27 million
1000 B.C.	50 million
500 B.C.	100 million

Social evolution of humans

Period years ago	Society type	Number of individuals
100,000-10,000	Bands	10s-100s
10,000-5000	tribes	100s-1,000s
5,000-3,000	Chiefdoms	1,000s-10,000s
3,000-1,000	States	10,000s-100,000s
1,000-Present	Empires	100,000-1,000,000s

• Traditions and Encounters textbook, page 24

The population nearly doubles every millennium as humans get better with agriculture and then from 1000 B.C. to 500 B.C. (only 500 years as compared to the 1000 years of the previous set) the population doubles in half the time. Agriculture also changed economics entirely. When humans were still nomadic hunter gatherers they had no time during the day to teach their children things other than how to hunt and gather. Finally, when the people could settle in villages and have a steady source of food do they really have time to take up special labor, "The concentration of large numbers of people in villages encouraged specialization of labor" (Traditions and Encounters textbook, p. 24).

The best known Neolithic settlement that demonstrates this is Çatal Hüyük. At this Neolithic village archeologists have found "pots, baskets, textiles, leather, stone and metal, wood carvings, carpets, beads, and jewelry..." (Traditions and Encounters textbook, p. 24). Pottery, metallurgy, and textile production were the main three industries that started at that time. After people started to settle and specialize and have more children the egalitarian society of the hunter gatherers also changed. Social classes emerged as familial wealth gets passed from one generation to the next. At Çatal Hüyük archeologists have found differences in interior decoration of one house to another, which clearly shows one family's wealth as compared to another's.

Subsequent Revolutions

Andrew Sherratt has argued that following upon the Neolithic Revolution was a second phase of discovery that he refers to as the secondary products revolution. Animals, it appears were first domesticated purely as a source of meat. The Secondary Products Revolution occurred when it was recognised that animals also provided a number of other useful products. These included:

- hides and skins (from no domesticated animals)
- manure for soil conditioning (from all domesticated animals)
- wool (from sheep, llamas, alpacas, and Angora goats)

- milk (from goats, cattle, yaks, sheep, horses and camels)
- traction (from oxen, onagers, donkeys, horses and camels)

Sherratt argues that this phase in agricultural development enabled humans to make use of the energy possibilities of their animals in new ways, and permitted permanent intensive subsistence farming and crop production, and the opening up heavier soils for farming. It also made possible nomadic pastoralism in semi arid areas, along the margins of deserts, and eventually led to the domestication of both the dromedary and bactrian camel. Overgrazing of these areas, particularly by herds of goats, greatly extended the areal extent of deserts. Living in one spot would have more easily permitted the accrual of personal possessions and an attachment to certain areas of land. From such a position, it is argued, prehistoric people were able to stockpile food to survive lean times and trade unwanted surpluses with others. Once trade and a secure food supply were established, populations could grow, and society would have diversified into food producers and artisans, who could afford to develop their trade by virtue of the free time they enjoyed because of a surplus of food. The artisans, in turn, were able to develop technology such as metal weapons. Such relative complexity would have required some form of social organisation to work efficiently and so it is likely that populations which had such organisation, perhaps such as that provided by religion were better prepared and more successful. In addition, the denser populations could form and support legions of professional soldiers. Also, during this time property ownership became increasingly important to all people. Ultimately, Childe argued that this growing social complexity, all rooted in the original decision to settle, led to a second Urban Revolution in which the first cities were built.

Disease

Throughout the development of sedentary societies, disease spread more rapidly than it had during the time in which hunter-gatherer societies existed. Inadequate sanitary practices and the domestication of animals may explain the rise in deaths and sickness during the Neolithic Revolution from disease, as diseases jumped from the animal to the human population. Some examples of diseases spread from animals to humans are influenza, smallpox, and measles. In concordance with a process of natural selection, the humans who first domesticated the big mammals quickly built up immunities to the diseases as within each generation the individuals with better immunities had better chances of survival.

In their approximately 10,000 years of shared proximity with animals, Eurasians and Africans became more resistant to those diseases compared with the indigenous populations encountered outside Eurasia and Africa. For instance, the population of most Caribbean and several Pacific Islands have been completely wiped out by diseases. According to the Population history of American indigenous peoples, 90% of the population of certain regions of North and South America were wiped out long before direct contact with Europeans. Some cultures like the Inca Empire did have one big mammal domesticated, the Llama, but the Inca did not drink its milk or live in a closed space with their herds, hence limiting the risk of contagion.

The causal link between the type or lack of agricultural development, disease and colonisation is not supported by colonization in other parts of the world. Disease increased after the establishment of British Colonial rule in Africa and India despite the areas having diseases that Europeans had no natural immunity to. In India agriculture developed during the Neolithic period with a wide range of animals domesticated. During colonial rule an estimated 23 million people died from cholera between 1865 and 1949, and millions more died from plague, malaria, influenza and tuberculosis.

In Africa European colonisation was accompanied by great epidemics, including malaria and sleeping sickness and despite parts of colonised Africa having little or no agriculture Europeans were more susceptible than these Africans. The increase of disease has been attributed to increased mobility of people, increased population density, urbanisation, environmental deterioration and irrigation schemes that helped to spread malaria rather than the development of agriculture.

Technology

In his book *Guns, Germs, and Steel*, Jared Diamond argues that Europeans and East Asians benefited from an advantageous geographical location which afforded them a head start in the Neolithic Revolution. Both shared the temperate climate ideal for the first agricultural settings, both were near a number of easily domesticable plant and animal species, and both were safer from attacks of other people than civilizations in the middle part of the Eurasian continent. Being among the first to adopt agriculture and sedentary lifestyles, and neighboring other early agricultural societies with whom they could compete and trade, both Europeans and East Asians were also among the first to benefit from technologies such as firearms and steel swords. In addition, they developed resistances to infectious disease, such as smallpox, due to their close relationship with domesticated animals. Groups of people who had not lived in proximity with other large mammals, such as the Australian Aborigines and American indigenous peoples were more vulnerable to infection and largely wiped out by diseases.

During and after the Age of Discovery, European explorers, such as the Spanish conquistadors, encountered other groups of people who had never or only recently adopted agriculture, such as in the Pacific Islands, or lacked domesticated big mammals such as the highlands people of Papua New Guinea.

MUSLIM AGRICULTURAL REVOLUTION

The Islamic Golden Age from the 8th century to the 13th century witnessed a fundamental transformation in agriculture known as the Arab Agricultural Revolution, Medieval Green Revolution, or Muslim Agricultural Revolution. The global economy established by Arab and other Muslim traders across the Old World, enabled the diffusion of many crops and farming techniques among different parts of the Islamic world, as well as the adaptation of crops and techniques from and to regions beyond the Islamic world. Crops from Africa such as sorghum, crops from China such as citrus fruits, and numerous crops from India such as mangos, rice, cotton and sugar cane, were distributed throughout Islamic lands, which previously had not grown these crops. Some writers have referred to the diffusion of numerous crops during this period as the Globalization of crops. These introductions, along with an increased mechanization of agriculture, led to major changes in economy, population distribution, vegetation cover, agricultural production and income, population levels, urban growth, the distribution of the labour force, linked industries, cooking, diet and clothing in the Islamic world.

Age of Discovery

The earliest forms of globalization began emerging during the Islamic Empire and the Islamic Golden Age, when the knowledge, trade and economies from many previously isolated regions and civilizations began integrating due to contacts with Muslim explorers, sailors, scholars, traders, and travelers. Some have called this period the "Pax Islamica" or "Afro-Asiatic age of discovery", in reference to the Jewish (known as Radhanites), Sogdian and Muslim Southwest Asian and North African traders and explorers who travelled most of the Old World, and established an early global economy across most of Asia and Africa and much of Europe, with their trade networks extending from the Atlantic Ocean and Mediterranean Sea in the west to the Indian Ocean and China Sea in the east. This helped establish the Islamic Empire (including the Rashidun, Umayyad, Abbasid and Fatimid caliphates) as the world's leading extensive economic power throughout the 7th-13th centuries. Several contemporary medieval Arabic reports also suggest that Muslim explorers from al-Andalus and the Maghreb may have travelled in expeditions across the Atlantic Ocean, possibly even to the Americas, between the 9th and 14th centuries.

Agricultural Innovations

Muslims widely practiced cash cropping and the modern crop rotation system where land was cropped four or more times in a two-year period. Winter crops were followed by summer ones, and in some cases there was in between. In areas where plants of shorter growing season were used, such as spinach and eggplants, the land could be cropped three or more times a year. In parts of Yemen, wheat yielded two harvests a year on the same land, as did rice in Iraq. Muslims developed a scientific approach based on three major elements; sophisticated systems of crop rotation, highly developed irrigation techniques, and the introduction of a large variety of crops which were studied and catalogued according to the season, type of land and amount of water they require. Numerous encyclopaedias on farming and botany were produced, with highly accurate precision and details.

Advanced Agricultural Systems

As early as the 9th century, an essentially modern agricultural system became central to economic life and organization in the Arab caliphates, replacing the largely export driven Roman model. Cities of the Near East, North Africa, and Moorish Spain were supported by elaborate agricultural systems which included extensive irrigation based on knowledge of hydraulic and hydrostatic principles, some of which were continued from Roman times. In later centuries, Persian Muslims began to function as a conduit, transmitting cultural elements, including advanced agricultural techniques, into Turkic lands and western India. The Muslims introduced what was to become an agricultural revolution based on four key areas:

- Development of a sophisticated system of irrigation using machines such as norias, water mills, water raising machines, dams and reservoirs. With such technology they managed to greatly expand the exploitable land area.
- The adoption of a scientific approach to farming enabled them to improve farming techniques derived from the collection and collation of relevant information throughout the whole of the known world. Farming manuals were produced in every corner of the Muslim world detailing where, when and how to plant and grow various crops. Advanced scientific techniques allowed leaders like Ibn al-Baitar to introduce new crops and breeds and strains of livestock into areas where they were previously unknown.

- Incentives based on a new approach to land ownership and labourers' rights, combining the recognition of private ownership and the rewarding of cultivators with a harvest share commensurate with their efforts. Their counterparts in Europe struggled under a feudal system in which they were almost slaves (serfs) with little hope of improving their lot by hard work.
- The introduction of new crops transforming private farming into a new global industry exported everywhere, including Europe, where farming was mostly restricted to wheat strains obtained much earlier via central Asia. Spain received what she in turn transmitted to the rest of Europe; many agricultural and fruit-growing processes, together with many new plants, fruit and vegetables. These new crops included sugar cane, rice, citrus fruit, apricots, cotton, artichokes, aubergines, and saffron. Others, previously known, were further developed. Muslims also brought to that country lemons, oranges, cotton, almonds, figs and sub-tropical crops such as bananas and sugar cane. Several were later exported from Spanish coastal areas to the Spanish colonies in the New World. Also transmitted via Muslim influence, a silk industry flourished, flax was cultivated and linen exported, and esparto grass, which grew wild in the more arid parts, was collected and turned into various articles.

Economic and Social Reforms

The caliphate understood that incentives were needed to increase productivity and wealth, thus enhancing tax revenues. Hence, they introduced a social transformation through the changed ownership of land, where any individual of any gender or any ethnic or religious background had the right to buy, sell, mortgage and inherit land for farming or any other purposes. They also introduced the signing of a contract for every major financial transaction concerning agriculture, industry, commerce, and employment. Copies of the contract was usually kept by both parties involved. The two types of economic systems that prompted agricultural development in the Islamic world were either politically-driven, by the conscious decisions of the central authority to develop under-exploited lands; or market-driven, involving the spread of advice, education, and free seeds, and the introduction of high value crops or animals to areas where they were previously unknown.

These led to increased subsistence, a high level of economic security that ensured wealth for all citizens, and a higher quality of life due to the introduction of artichokes, spinach, aubergines, carrots, sugar cane, and various exotic plants; vegetables being available all year round without the need to dry them for winter; citrus and olive plantations becoming a common sight, market gardens and orchards springing up in every Muslim city; intense cropping and the technique of intensive irrigation agriculture with land fertility replacement; a major increase in animal husbandry; higher quality of wool and other clothing materials; and the introduction of selective breeding of animals from different parts of the Old World resulting in improved horse stocks and the best load-carrying camels.

The average life expectancy in the lands under Islamic rule also experienced an increase, due to the Agricultural Revolution as well as improved medical care. In contrast to the average lifespan in the ancient Greco-Roman world (22-28 years), the average lifespan in the early Islamic Caliphate was more than 35 years. The average lifespans of the Islamic scholarly class in particular was much higher: 84.3 years in 10th-11th century Iraq and Persia, 72.8 years in the 11th century Middle East, 69-75 years in 11th century Islamic Spain, 75 years in 12th century Persia, and 59-72 years in 13th century Persia. The Islamic Empire also experienced a growth in literacy, having the highest literacy rate of the Middle Ages, comparable to Athens' literacy in Classical Antiquity but on a larger scale.

Sugar Industry

During the Muslim Agricultural Revolution, sugar production was refined and transformed into a large-scale industry by the Arabs. The Arabs and Berbers diffused sugar throughout the Arab Empire from the 8th century.

Other Innovations

Many other agricultural innovations were introduced by Muslim farmers and engineers, such as new forms of land tenure, improvements in irrigation, a variety of sophisticated irrigation methods, the introduction of fertilizers and widespread artificial irrigation systems, the development of gravity-flow irrigation systems from rivers and springs, the first uses of noria and chain pumps for irrigation purposes, the establishment of the sugar cane industry in the Mediterranean and experimentation in sugar cultivation, numerous advances in industrial milling and water-raising machines (see Industrial growth below), and many other improvements and innovations.

Agricultural Sciences

During the Muslim Agricultural Revolution, Muslim scientists laid the foundations of agricultural science, which included significant advances in the fields of agronomy, astronomy, botany, earth science, environmental philosophy, and environmental science. In particular, the experimental scientific method was introduced into the field in the 13th century by the Andalusian-Arab botanist Abu al-Abbas al-Nabati, the teacher of Ibn al-Baitar. Al-Nabati introduced empirical techniques in the testing, description and identification of numerous materia medica, and he separated unverified reports from those supported by actual tests and observations.

The earliest known work dedicated to the study of agriculture was Ibn Wahshiyya's *Nabatean Agriculture*, which also dealt with the related field of botany and was also an early cookbook. The early Arab lexicographs were the first known works to separate the two disciplines of agriculture and botany, though both were considered part of the medical sciences due to agriculture's primary role being to feed and botany's primary role being to heal. The agricultural sciences were known by the Arabic term *filaha*, which had a dual-meaning, to both care for the Earth and to take care of plants. Many of the early Islamic authors on botany were often philologists, due to their role in the translation of ancient scientific texts. This was also the case with early Arabic zoology, like with al-Jahiz for example.

Al-Asma'i was the earliest known Arab biologist, botanist and zoologist; his works include the *Book of Distinction*, *Book of the Wild Animals*, *Book of the Horse*, and *Book of the Sheep*.

Agronomy

Muslim agriculturists demonstrated advanced agronomic, agrotechnical and economic knowledge in areas such as meteorology, climatology, hydrology, soil occupation, and the economy and management of agricultural enterprises. They also demonstrated agricultural knowledge in areas such as pedology, agricultural ecology, irrigation, preparation of soil, planting, spreading of manure, killing herbs, sowing, cutting trees, grafting, pruning vine], prophylaxis, phytotherapy, the care and improvement of microbiological cultures and plants, and the harvest and storage of crops.

Ibn Wahshiyya's *Nabatean Agriculture* was an early Arabic work on agronomy and agriculture. The following eight chapters of the book are dedicated to water in the context of agriculture:

1. Research of water and related technical knowledge
2. Digging wells and increasing their flow using proven artifices and techniques
3. The drilling of wells
4. Artifices used to increase water in a well
5. Making water rise up a very deep well
6. Augmenting the quantity of water in wells and sources
7. Modifying and improving the taste of water
8. "On the difference in nature and action of the water according to its position" close of far away "with regard to the ecliptic"

The *Nabatean Agriculture* then goes on to discuss a number of other complex issues on agriculture, including the management of an agricultural enterprise and the duties of the owner regarding his enterprise and workers; the official (*wakil*) in charge of the management of the enterprise, his obligation towards the farmer]s, and applying the instructions he receives from his boss; the weather forecasting of atmospheric changes and signs from the planetary astral alterations; signs of rain based on observation of the lunar phases, nature of thunder and lightning, direction of sunrise, behaviour of certain plants and animals, and weather forecasts based on the movement of winds; the recognition of plant tissue cultures which succeed in certain years; a list of work to be done in each month of year; the position of the moon relative to the Earth; the required knowledge of a farmer and the owner of an agricultural enterprise; pollenized air and winds; and formation of winds and vapours.

Other agricultural topics discussed in the *Nabatean Agriculture* include the causes of the corruption of plants and of torrential rain; the nature of soils and their different flavours; the manure; how to get rid of bad herbs and how to cut plants which need to be cut; and a number of other agricultural topics.

In 12th century al-Andalus, Ibn al-'Awwam al-Ishbili wrote the *Kitab al-Filaha* which synthesized his own agricultural knowledge with that of the *Nabatean Agriculture* and his other Arabic predecessors. This work also described 585 microbiological cultures, 55 of which concern fruit trees. This work was influential in Europe after it was translated into Spanish by Banqueri in Madrid in 1801 and into French by Clement-Mullet in Paris in 1864.

Astronomy and Meteorology

Another innovation during this period was the application of astronomy to agriculture and botany. As weather forecasting predictions and the measurement of time and the onset of seasons became more precise and reliable, farmers became informed of these advances and often employed them in agriculture. They also benefited from the compilation of calendars with information on when to plant each type of crop, when to graft trees, when and how to fertilize crops, when to harvest, and what to eat and what to avoid at each time of year. These advances made it possible for farmers to plan the growth of each of their crops for specific markets and at specific times of the year. Parts of al-Dinawari's *Book of Plants* deal with the applications of astronomy and meteorology to agriculture. It describes the astronomical and meteorological character of the sky, the planets and constellations, the sun and moon, the lunar phases indicating seasons and rain, the *anwa* (heavenly bodies of rain), and atmospheric phenomena such as winds, thunder, lightning, snow, floods, valleys, rivers, lakes, wells and other sources of water.

Botany

Muslims developed a scientific approach to botany and agriculture based on three major elements: sophisticated systems of crop rotation, highly developed irrigation techniques, and the introduction of a large variety of crops which were studied and catalogued according to the season, type of land and amount of water they require. Numerous encyclopaedias on botany were produced, with highly accurate precision and details.

The 9th century botanist al-Dinawari is considered the founder of Arabic botany. He wrote a botanical encyclopedia entitled *Kitab al-Nabat* (*Book of Plants*), which consisted of six volumes. Only the third and fifth volumes have survived, though the sixth volume has partly been reconstructed based on citations from later works. In the surviving portions of his works, 637 plants are described from the letters *sin* to *ya*. He also discusses plant evolution from its birth to its death, describing the phases of plant growth and the production of flowers and fruit.

In the early 13th century, Ibn al-Baitar published the *Kitab al-Jami fi al-Adwiya al-Mufrada*, which is considered one of the greatest botanical compilations and pharmaceutical encyclopedias, and was a botanical authority for centuries. It contains details on at least 1,400 different plants, foods, and drugs, 300 of which were his own original discoveries. The *Kitab al-Jami fi al-Adwiya al-Mufrada* was also influential in Europe after it was translated into Latin in 1758, where it was being used up until the early 19th century.

Earth Science

Muslim scientists made a number of contributions to the earth sciences. Alkindus introduced experimentation into the

Earth sciences. Parts of al-Dinawari's *Book of Plants* deals with the Earth sciences in the context of agriculture. He considers the Earth, stone and sands, and describes different types of ground, indicating which types are more convenient for plants and the qualities and properties of good ground.

Biruni is considered the father of geodesy for his important contributions to the field, along with his significant contributions to geography and geology. Among his writings on geology, Biruni wrote the following on the geology of India:

> *"But if you see the soil of India with your own eyes and meditate on its nature, if you consider the rounded stones found in earth however deeply you dig, stones that are huge near the mountains and where the rivers have a violent current: stones that are of smaller size at a greater distance from the mountains and where the streams flow more slowly: stones that appear pulverised in the shape of sand where the streams begin to stagnate near their mouths and near the sea-if you consider all this you can scarcely help thinking that India was once a sea, which by degrees has been filled up by the alluvium of the streams."*

John J. O'Connor and Edmund F. Robertson write in the *MacTutor History of Mathematics archive*:

> *"Important contributions to geodesy and geography were also made by al-Biruni. He introduced techniques to measure the earth and distances on it using triangulation. He found the radius of the earth to be 6339.6 km, a value not obtained in the West until the 16th century. His* Masudic canon *contains a table giving the coordinates of six hundred places, almost all of which he had direct knowledge."*

In geology, Avicenna hypothesized on two causes of mountains in *The Book of Healing*. In cartography, the Piri Reis map drawn by the Ottoman cartographer Piri Reis in 1513, was one of the earliest world maps to include the Americas, and perhaps the first to include Antarctica. His map of the world was considered the most accurate in the 16th century.

Environmental Philosophy

Perhaps due to resource scarcity in most Islamic nations, there was an emphasis on limited (and some claim also sustainable) use of natural capital, i.e. producing land.

Traditions of haram and hima and early urban planning were expressions of strong social obligations to stay within carrying capacity and to preserve the natural environment as an obligation of khalifa or "stewardship".

Muhammad is considered a pioneer of environmentalism for his teachings on environmental preservation. His hadiths on agriculture and environmental philosophy were compiled in the "Book of Agriculture" of the *Sahih Bukhari*, which included the following saying:

> *"There is none amongst the believers who plants a tree, or sows a seed, and then a bird, or a person, or an animal eats thereof, but it is regarded as having given a charitable gift [for which there is great recompense]."*

Several such statements concerning the environment are also found in the Qur'an, such as the following:

> *"And there is no animal in the earth nor bird that flies with its two wings, but that they are communities like yourselves."*

Environmental Science

The earliest known treatises dealing with environmentalism and environmental science, especially pollution, were Arabic medical treatises written by al-Kindi, Qusta ibn Luqa, al-Razi, Ibn Al-Jazzar, al-Tamimi, al-Masihi, Avicenna, Ali ibn Ridwan, Ibn Jumay, Isaac Israeli ben Solomon, Abd-el-latif, Ibn al-Quff, and Ibn al-Nafis.

Their works covered a number of subjects related to pollution such as air pollution, water pollution, soil contamination, municipal solid waste mishandling, and environmental impact assessments of certain localities. Cordoba, al-Andalus also had the first waste containers and waste disposal facilities for litter collection.

Zoology

In the zoology field of biology, Muslim biologists developed theories on evolution and natural selection which were widely taught in medieval Islamic schools. John William Draper, a contemporary of Charles Darwin, considered the "Mohammedan theory of evolution" to be developed "much farther than we are disposed to do, extending them even to inorganic or mineral things." According to al-Khazini, ideas on evolution were widespread among "common people" in the Islamic world by the 12th century. The first Muslim biologist to develop a theory on evolution was al-Jahiz (781-869). He wrote on the effects of the environment on the likelihood of an animal to survive, and he first described the struggle for existence and an early form of natural selection. Al-Jahiz was also the first to discuss food chains, and was also an early adherent of environmental determinism, arguing that the environment can determine the physical characteristics of the inhabitants of a certain community and that the origins of different human skin colors is the result of the environment.

Ibn al-Haytham wrote a book in which he argued for evolutionism (although not natural selection), and numerous other Islamic scholars and scientists, such as Ibn Miskawayh, the Brethren of Purity, al-Khazini, Abû Rayhân al-Bîrûnî, Nasir al-Din Tusi, and Ibn Khaldun, discussed and developed these ideas. Translated into Latin, these works began to appear in the West after the Renaissance and appear to have had an impact on Western science.

Ibn Miskawayh's *al-Fawz al-Asghar* and the Brethren of Purity's *Encyclopedia of the Brethren of Purity* (*The Epistles of Ikhwan al-Safa*) expressed evolutionary ideas on how species evolved from matter, into vapor, and then water, then minerals, then plants, then animals, then apes, and then humans. These works were known in Europe and likely had an influence on Darwinism.

Capitalist Market Economy

Capitalism developed much earlier in Islamic regions than in the Occident. Subhi Y. Labib argues the reason for this was the growing trade economy of the Muslim world, and security from Barbarian invasions. The first market economy and earliest forms of merchant capitalism took root between the 8th–12th centuries in the Caliphate, which are referred to as "Islamic capitalism". A vigorous monetary economy was created on the basis of the expanding levels of circulation of a stable high-value currency (the dinar) and the integration of monetary areas that were previously independent.

Innovative new business techniques and forms of business organisation were introduced by economists, merchants and traders during this time. Such innovations included the earliest trading companies, big businesses, contracts, bills of exchange, long-distance international trade, the first forms of partnership (*mufawada*) such as limited partnerships (*mudaraba*), and the earliest forms of credit, debt, profit, loss, capital (*al-mal*), capital accumulation (*nama al-mal*), circulating capital, capital expenditure, revenue, cheques, promissory notes, trusts and charitable trusts (see *Waqf*), startup companies, savings accounts, transactional accounts, pawning, loaning, exchange rates, bankers, money changers, ledgers, deposits, assignments, the double-entry bookkeeping system, and lawsuits.

Organizational enterprises similar to corporations independent from the state also existed in the medieval Islamic world, while the agency and aval institutions (see *Hawala*) was also introduced. Many of these early capitalist concepts were adopted and further advanced in medieval Europe from the 13th century onwards.

The systems of contract relied upon by merchants was very effective. Merchants would buy and sell on commission, with money loaned to them by wealthy investors, or a joint investment of several merchants, who were often Muslim, Christian and Jewish. Recently, a collection of documents was found in an Egyptian synagogue shedding a very detailed and human light on the life of medieval Middle Eastern merchants. Business partnerships would be made for many commercial ventures, and bonds of kinship enabled trade networks to form over huge distances.

Crops

Hundreds of crops were diffused throughout the Islamic world and beyond as a result of the Muslim Agricultural Revolution, some of which include artichokes, bananas, coconut palms, colocasia, cotton, eggplants, hard wheat, lemons, limes, mangos, plantains, rice, sorghum, sour oranges, spinach, sugar cane, and watermelons, among hundreds of other crops.

Industrial Growth

Muslim engineers in the Islamic world were responsible for numerous innovative industrial uses of hydropower, early industrial uses of tidal power, wind power, and fossil fuels such as petroleum, and the earliest large factory complexes (*tiraz* in Arabic). The industrial uses of watermills in the Islamic world date back to the 7th century, while horizontal-wheeled and vertical-wheeled water mills were both in widespread use since at least the 9th century. A variety of industrial mills were first invented in the Islamic world, including fulling mills, gristmills, hullers, paper mills, sawmills, shipmills, stamp mills, steel mills, sugar mills, tide mills, and windmills. By the 11th century, every province throughout the Islamic world had these industrial mills in operation, from al-Andalus and North Africa to the Middle East and Central Asia. Muslim engineers also invented crankshafts and water turbines, first employed gears in mills and water-raising machines, and pioneered the use of dams as a source of water power, used to provide additional power to watermills and water-raising machines. Such advances made it possible for many industrial tasks that were previously driven by manual labour in ancient times to be mechanized and driven by machinery instead in the medieval Islamic world. The transfer of these technologies to medieval Europe later laid the foundations for the Industrial Revolution in 18th century Europe.

Many industries were generated due to the Muslim Agricultural Revolution, including the earliest industries for agribusiness, astronomical instruments, ceramics, chemicals, distillation technologies, clocks, glass, mechanical hydropowered and wind powered machinery, matting, mosaics, [[Pulp and paper industry | pulp and paper, perfumery, petroleum, pharmaceuticals, rope-making, shipping, shipbuilding, silk, sugar, textiles, weapons, and the mining of minerals such as sulfur, ammonia, lead and iron]. The first large factory complexes (*tiraz*) were built for many of these industries. Knowledge of these industries were later transmitted to medieval Europe, especially during the Latin translations of the 12th century, as well as before and after. For example, the first glass factories in Europe were founded in the 11th century by Egyptian craftsmen in Greece. The agricultural and handicraft industries also experienced high levels of growth during this period.

Industrial Milling

Muslim engineers pioneered two solutions to achieve the maximum output from a water mill. The first solution was to mount them to piers of bridges to take advantage of the increased flow. The second solution was the shipmill, a unique type of water mill powered by water wheels mounted on the sides of ships moored in midstream. This was first employed along the Tigris and Euphrates rivers in 10th century Iraq, where large shipmills made of teak and iron could produce 10 tons of flour from corn every day for the granary in Baghdad. Industrial water mills were also employed in the first large factory complexes built in al-Andalus between the 11th and 13th centuries.

Windmills were first built in Sistan, Afghanistan, from the 7th century. These were vertical axle windmills, which had long vertical driveshafts with rectangle shaped blades. The first windmill was built by the Rashidun caliph Umar (634–44). Made of six to twelve sails covered in reed matting or cloth material, these windmills were used to grind corn and draw up water, and were used in the gristmilling and sugarcane industries.

After paper was introduced into the Islamic world by Chinese prisoners following the Battle of Talas, Muslims made significant improvements to papermaking and built the first paper mills in Baghdad, Iraq, as early as 794. Papermaking was transformed from an art into a major industry as a result.

The first fulling mills were later invented in the 10th century, followed by the first stamp mills and steel mills in the 11th century.

The first gristmills were invented by Muslim engineers in the Islamic world, and were used for grinding corn and other seeds to produce meals, and many other industrial uses such as fulling cloth, husking rice, papermaking, pulping sugarcane, and crushing metallic ores before extraction. Gristmills in the Islamic world were often made from both watermills and windmills. In order to adapt water wheels for gristmilling purposes, cams were used for raising and releasing trip hammers to fall on a material. The first water turbine, which had water wheels with curved blades onto which water flow was directed axially, was also first invented in the Islamic world, and was described in a 9th century Arabic text for use in a watermill.

Labour Force

The labor force in the caliphate were employed from diverse ethnic and religious backgrounds, while both men and women were involved in diverse occupations and economic activities. Women were employed in a wide range of commercial activities and diverse occupations in the primary sector (as farmers for example), secondary sector (as construction workers, dyers, spinners, etc.) and tertiary sector (as investors, doctors, nurses, presidents of guilds, brokers, peddlers, lenders, scholars, etc.). Muslim women also held a monopoly over certain branches of the textile industry, the largest and most specialized and market-oriented industry at the time, in occupations such as spinning, dying, and embroidery. In comparison, female property rights and wage labour were relatively uncommon in Europe until the Industrial Revolution in the 18th and 19th centuries.

The division of labour was diverse and had been evolving over the centuries. During the 8th–11th centuries, there were on average 63 unique occupations in the primary sector of economic activity (extractive), 697 unique occupations in the secondary sector (manufacturing), and 736 unique occupations in the tertiary sector (service). By the 12th century, the number of unique occupations in the primary sector and secondary sector decreased to 35 and 679 respectively, while the number of unique occupations in the tertiary sector increased to 1,175. These changes in the division of labour reflect the increased mechanization and use of machinery to replace manual labour and the increased standard of living and quality of life of most citizens in the Caliphate.

An economic transition occurred during this period, due to the diversity of the service sector being far greater than any other previous or contemporary society, and the high degree of economic integration between the labour force and the economy. Islamic society also experienced a change in attitude towards manual labour. In previous civilizations such as ancient Greece and in contemporary civilizations such as early medieval Europe, intellectuals saw manual labour in a negative light and looked down on them with contempt. This resulted in technological stagnation as they did not see the need for machinery to replace manual labour. In the Islamic world, however, manual labour was seen in a far more positive light, as intellectuals such as the Brethren of Purity likened them to a participant in the act of creation, while Ibn Khaldun alluded to the benefits of manual labour to the progress of society.

Mechanical Technology

The reciprocating suction piston pump with valves and crankshaft-connecting rod mechanism used for raising water, invented by al-Jazari, the father of modern day engineering.

Noria and chain pump (saqiya) machines became more widespread during the Muslim Agricultural Revolution, when Muslim engineers made a number of improvements to the device. These include the first uses of noria and chain pumps for irrigation purposes, and the invention of the flywheel mechanism, used to smooth out the delivery of power from a driving device to a driven machine, which was first invented by Ibn Bassal (fl. 1038–75) of al-Andalus, who pioneered the use of the flywheel in the saqiya and noria.

In 1206, al-Jazari invented a variety of machines for raising water, which were the most efficient in his time, as well as water wheels with cams on their axle used to operate automata. He invented the crankshaft and connecting rod, and employed them in a crank-connecting rod system for two of these water-raising machines. His invention of the crankshaft is considered the most important single mechanical invention after the wheel, as it transforms continuous rotary motion into a linear reciprocating motion, and is central to modern machinery such as the steam engine and the internal combustion engine.

Urbanization

There was a significant increase in urbanization during this period, due to numerous scientific advances in fields such as agriculture, hygiene, sanitation, astronomy, medicine and engineering. This also resulted in a rising middle class population. As urbanization increased, Muslim cities grew unregulated, resulting in narrow winding city streets and neighborhoods separated by different ethnic backgrounds and religious affiliations. These qualities proved efficient for transporting goods to and from major commercial centers while preserving the privacy valued by Islamic family life. Suburbs lay just outside the walled city, from wealthy residential communities, to working class semi-slums. City garbage dumps were located far from the city, as were clearly defined cemeteries which were often homes for criminals. A place of prayer was found just near one of the main gates, for religious festivals and public executions. Similarly, Military Training grounds were found near a main gate.

While varying in appearance due to climate and prior local traditions, Islamic cities were almost always dominated by a merchant middle class. Some peoples' loyalty towards their neighborhood was very strong, reflecting ethnicity and religion, while a sense of citizenship was at times uncommon (but not in every case). The extended family provided the foundation for social programs, business deals, and negotiations with authorities. Part of this economic and social unit were often the tenants of a wealthy landlord.

State power normally focused on Dar al Imara, the governor's office in the citadel. These fortresses towered high above the city built on thousands of years of human settlement. The primary function of the city governor was to provide for defence and to maintain legal order. This system would be responsible for a mixture of autocracy and autonomy within the city. Each neighborhood, and many of the large tenement blocks, elected a representative to deal with urban authorities. These neighborhoods were also expected to organize their young men into a militia providing for protection of their own neighborhoods, and as aid to the professional armies defending the city as a whole.

The head of the family was given the position of authority in his household, although a qadi, or judge was able to negotiate and resolve differences in issues of disagreements within families and between them. The two senior representatives of municipal authority were the qadi and the muhtasib, who held the responsibilities of many issues, including quality of water, maintenance of city streets, containing outbreaks of disease, supervising the markets, and a prompt burial of the dead.

Another aspect of Islamic urban life was waqf, a religious charity directly dealing with the qadi and religious leaders. Through donations, the waqf owned many of the public baths and factories, using the revenue to fund education, and to provide irrigation for orchards outside the city. Following expansion, this system was introduced into Eastern Europe by Ottoman Turks.

While religious foundations of all faiths were tax exempt in the Muslim world, civilians paid their taxes to the urban authorities, soldiers to the superior officer, and landowners to the state treasury. Taxes were also levied on an unmarried man until he was wed. Instead of zakat, the mandatory charity required of Muslims, non-Muslims were required to pay the jizya, a discriminatory religious tax, imposed on Christians and Jews. During the Muslim Conquests of the 7th and 8th centuries conquered populations were given the three choices of either converting to Islam, paying the jizya, or dying by the sword.

Animals brought to the city for slaughter were restricted to areas outside the city, as were any other industries seen as unclean. The more valuable a good was, the closer its market was to the center of town. Because of this, booksellers and goldsmiths clustered around the main mosque at the heart of the city.

Civil Engineering

Many dams, acequia and qanat water supply systems, and "Tribunal of Waters" irrigation systems, were built during the Islamic Golden Age and are still in use today in the Islamic world and in formerly Islamic regions of Europe such as Sicily and the Iberian Peninsula, particularly in the Andalusia, Aragon] and Valencia provinces of Spain. The Arabic systems of irrigation and water distribution were later adopted in the Canary Islands and Americas due to the Spanish and are still used in places like Texas, Mexico, Peru, and Chile.

Muslim cities also had advanced domestic water systems with sewers, public baths, drinking fountains, piped drinking water supplies, and widespread private and public toilet and bathing facilities. By the 10th century, Cordoba had 700 mosques, 60,000 palaces, and 70 libraries, the largest of which had 600,000 books, while as many as 60,000 treatises, poems, polemics and compilations were published each year in al-Andalus. The library of Cairo had more than 100,000 books, while the library of Tripoli is said to have had as many as three million books. The number of important and original Arabic works on science that have survived is much larger than the combined total of Greek and Latin works on science.

Islamic cities also had an early public health care service. "The extraordinary provision of public bath-houses, complex sanitary systems of drainage (more extensive even than the famous Roman infrastructures), fresh water supplies, and the large and sophisticated urban hospitals, all contributed to the general health of the population."

BRITISH AGRICULTURAL REVOLUTION

The British Agricultural Revolution describes a period of development in Britain between the 17th century and the end of the 19th century, which saw a massive increase in agricultural productivity and net output. This in turn supported unprecedented population growth, freeing up a significant percentage of the workforce, and thereby helped drive the Industrial Revolution. How this came about is not entirely clear. In recent decades, enclosure, mechanization, four-field crop rotation, and selective breeding have been highlighted as primary causes, with credit given to relatively few individuals.

Enclosure

Prior to the 18th century, agriculture had been much the same across Europe since the Middle Ages. The open field system was essentially post-feudal, with each farmer subsistence-cropping strips of land in one of three or four large fields held in common and splitting up the products likewise.

Beginning as early as the 12th century, some of the common fields in Britain were enclosed into individually owned fields, and the process rapidly accelerated in the 15th and 16th centuries. This led to farmers losing their land and their grazing rights, and left many unemployed. In the 16th and 17th centuries, the practice of enclosure was denounced by the Church, and legislation was drawn up against it; but the developments in agricultural mechanization during the 18th century required large, enclosed fields in order to be workable. This led to a series of government acts, culminating finally in the General Enclosure Act of 1801. While farmers received compensation for their strips, it was minimal, and the loss of rights for the rural population led to an increased dependency on the Poor law. Surveying and legal costs weighed heavily on poor farmers, who sometimes even had to sell their share of the land to pay for its being split up. Only a few found work in the (increasingly mechanised) enclosed farms. Most were forced to relocate to the cities to try to find work in the emerging factories of the Industrial Revolution. By the end of the 18th century the process of enclosure was complete.

Mechanization

Jethro Tull made early advancements in agricultural technology with his seed drill (1701) — a mechanical seeder which distributed seeds efficiently across a plot of land. However, he was not the first to invent a seed drill. It took a century after the publication in 1731 of his *Horse hoeing husbandry* for farmers to widely adopt the technology. This may sound to be obvious and necessary processes. But they were not practiced in Europe until the eighteenth century and he was the major contributor in this conversion.

The Chinese were doing this at least by the sixth century BC, and were thus a good 2,200 years in advance of the West in one of the most sensible aspects of agriculture. Contribution of Chinese agricultural methods and tools have been rarely mentioned, rather hidden or kept as minor.

The Chinese iron ploughs, which is generally called 'Dutch design' because the Dutch brought them from China, contributed drastic improvement in productivity not only in Britain but also in Europe and the U.S. Tull's seed drill was inspired by the excellent Chinese counterpart, which never arrived in Europe. The Chinese rotary winnowing fan revolutionized winnowing process since its introduction to Europe.

Joseph Foljambe's Rotherham plough of 1730, while not the first iron plough, was the first iron plough to have any commercial success, combining an earlier Dutch design with a number of technological innovations. Its fittings and coulter were made of iron and the mouldboard and share were covered with an iron plate making it lighter to pull and more controllable than previous ploughs.

It remained in use in Britain until the development of the tractor. It was followed by John Small of Doncaster and Berwickshire in 1763, whose 'Scots Plough' used an improved cast iron shape to turn the soil more effectively with less draft, wear, or strain on the ploughing team.

Andrew Meikle's threshing machine of 1786 was the final straw for many farm labourers, and led to the 1830 agricultural rebellion of Captain Swing (a probably mythical character comparable to the Luddite's Ned Ludd).

In the 1850s and '60s John Fowler, an agricultural engineer, pioneered the use of steam engines for ploughing and digging drainage channels.

Four-field Crop Rotation

During the Middle Ages, the open field system had employed a three year crop rotation, with a different crop in each of the three fields, eg. wheat and barley in two, with the third fallow. 'Fallow' is a term which means that the field is empty, there is nothing growing there. Over the following two centuries, the regular planting of nitrogen-rich legumes in the fields which were previously fallow slowly increased the fertility of croplands. The planting of legumes (leguminosae, plants of the pea/bean family) helped to increase plant growth in the empty field because they used a different set of nutrients to grow than the grains. The legumes put back nutrients the grains used, nitrates produced from nitrogen in the atmosphere, and the grains put back the minerals the legumes used. In a way, they fed each other. Other crops that were occasionally grown were flax, and members of the mustard family. Medieval record keepers did not distinguish between rape seed or other mustards grown for animal feed and mustard grown for mustard seed for condiments. When the pastures were brought back into crop production after their long fallow, their fertility was much greater than they had been in medieval times. The farmers in Flanders (current day Belgium), however, discovered a still more effective four-field rotation system, introducing turnips and clover to replace the fallow year. Clover was both an ideal fodder crop, and it actually improved grain yields in the following year (clover is part of the pea family, leguminosae). The improved grain production simultaneously increased livestock production. Farmers could grow more livestock because there was more food, and manure was an excellent fertilizer, so they could have even more productive crops. Charles Townshend learned the four field system from Flanders and introduced it to Great Britain in 1730.

Selective Breeding

In England, Robert Bakewell and Thomas Coke introduced selective breeding (mating together two animals with particularly desirable characteristics), and inbreeding (to stabilize certain qualities) in order to reduce genetic diversity in desirable animals programs from the mid 18th century.

Robert Bakewell cross-bred the Lincoln and Longhorn sheep to produce the New Leicester variety. These methods proved successful in the production of larger and more profitable livestock.

Technology from Flanders

The British Agricultural Revolution was sparked in part by advancements in Flanders, including the aforementioned four-crop rotation. Due to the large and dense population of Flanders, which forced farmers to take advantage of every inch of usable land, the country had become a pioneer in drainage and reclamation technology. Many Flemish experts went to The Netherlands and reclaimed land there. Finally, Dutch experts like Cornelius Vermuyden brought the technology to Britain.

Effects on History

The Agricultural Revolution in Britain proved to be a major turning point in history. The population in 1750 reached the level of 5.7 million. This had happened before: in around 1300 and again in 1650. Each time, the appropriate agricultural infrastructure to support a population this high was not present, and the population fell. However, by 1750, when the population reached this level again, an onset in agricultural technology and new methodology allowed the population growth to be sustained.

The increase in population led to more demand from the people for goods such as clothing. A new class of landless labourers, products of enclosure, provided the basis for cottage industry, a stepping stone to the Industrial Revolution. To supply continually growing demand, shrewd businessmen began to pioneer new technology to meet demand from the people. This led to the first industrial factories. People who once were farmers moved to large cities to get jobs in the factories. It should be noted that the British Agricultural Revolution not only made the population increase possible, but also increased the yield per agricultural worker, meaning that a larger percentage of the population could work in these new, post-Agricultural Revolution jobs.

The British Agricultural Revolution was the cause of drastic changes in the lives of British women. Before the Agricultural Revolution, women worked alongside their husbands in the fields and were an active part of farming. The increased efficiency of the new machinery, along with the fact that this new machinery was often heavier and difficult for a woman to wield, made this unnecessary and impractical, and women were relegated to other roles in society. To supplement the family's income, many went into cottage industries. Others became domestic servants or were forced into professions such as prostitution. The new, limited roles of women, dubbed by one historian as "this defamation of women workers", (Valenze) fueled prejudices of women only being fit to work in the home, and also effectively separated them from the new, mechanized areas of work, leading to a divide in the pay between men and women.

Towards the end of the 19th century, the substantial gains in British agricultural productivity were rapidly offset by competition from cheaper imports, made possible by advances in transportation, refrigeration, and many other technologies. From that point, farming in Britain entered a period of economic struggle which continues to the present day.

6

GREEN REVOLUTION

Green Revolution usually refers to the transformation of agriculture that began in 1945. One significant factor came at the request of the Mexican government to establish an agricultural research station to develop more varieties of wheat that could be used to feed the rapidly growing population of the country.

In 1943 Mexico imported half its wheat; in 1956, the Green Revolution had made Mexico self-sufficient; by 1964, Mexico exported half a million tons of wheat. The associated transformation has continued as the result of programs of agricultural research, extension, and infrastructural development, instigated and largely funded by the Rockefeller Foundation, along with the Ford Foundation and other major agencies.

Many agronomists state that the Green Revolution has allowed food production to keep pace with worldwide population growth while others state that it caused the great population increases seen today. The Green Revolution has had major social and ecological impacts, making it a popular topic of study among sociologists.

The term "Green Revolution" was first used in 1968 by former USAID director William Gaud, who noted the spread of the new technologies and said, "These and other developments in the field of agriculture contain the makings of a new revolution. It is not a violent Red Revolution like that of the Soviets, nor is it a White Revolution like that of the Shah of Iran. I call it the Green Revolution."

History

Indian Success

With the experience of agricultural development begun in Mexico by Norman Borlaug in 1943 judged as a success, the Rockefeller Foundation sought to spread the Green Revolution to other nations. The Office of Special Studies in Mexico became an informal international research institution in 1959, and in 1963 it formally became CIMMYT, The International Maize and Wheat Improvement Center.

In 1961 India was on the brink of mass famine. Norman Borlaug was invited to India by the adviser to the Indian minister of agriculture M. S. Swaminathan. Despite bureaucratic hurdles imposed by India's grain monopolies, the Ford Foundation and Indian government collaborated to import wheat seed from CIMMYT. Punjab was selected by the Indian government to be the first site to try the new crops because of its reliable water supply and a history of agricultural success. India began its own Green Revolution program of plant breeding, irrigation development, and financing of agrochemicals.

India soon adopted IR8-a rice semi-dwarf variety developed by the International Rice Research Institute(IRRI) that could produce more grains of rice per plant when grown properly with fertilizer and irrigation. In 1968, Indian agronomist S.K. De Datta published his findings that IR8 rice yielded about 5 tons per hectare with no fertilizer, and almost 10 tons per hectare under optimal conditions.

This was 10 times the yield of traditional rice. IR8 was a success throughout Asia, and dubbed the "Miracle Rice". In the 1960s, rice yields in India were about two tons per hectare; by the mid-1990s, they had risen to six tons per hectare. In the 1970s, rice cost about $550 a ton; in 2001, it cost less than $200 a ton. India became one of the world's most successful rice producers, and is now a major rice exporter, shipping nearly 4.5 million tons in 2006.

Famine in India, once accepted as inevitable, has not returned since the introduction of Green Revolution agriculture.

IR8 and the Philippines

In 1960, the Government of the Republic of the Philippines with Ford and Rockefeller Foundations established IRRI. A rice crossing between Dee-geo-woo-gen and Peta was done at IRRI in 1962. In 1966, one of the breeding lines became a new cultivar, IR8. IR8 required the use of fertilizers and pesticides, but produced substantially higher yields than the traditional cultivars. Annual rice production in the Philippines increased from 3.7 to 7.7 million tonnes in two decades. The switch to IR8 rice made the Philippines a rice exporter for the first time in the 20th century. At the same time, heavy pesticide use reduced the number of fish and frog species found in rice paddies.

CGIAR

In 1970, foundation officials proposed a worldwide network of agricultural research centers under a permanent secretariat. This was further supported and developed by the World Bank; on May 19, 1971, with the FAO, IFAD and UNDP as co-sponsors, the Consultative Group on International Agricultural Research was established. CGIAR, has added many research centers throughout the world.

CGIAR has responded, at least in part, to criticisms of Green Revolution methodologies. This began in the 1980s, and mainly was a result of pressure from donor organizations. Methods like Agroecosystem Analysis and Farming System Research have been adopted to gain a more holistic view of agriculture. Methods like Rapid Rural Appraisal and Participatory Rural Appraisal have been adopted to help scientists understand the problems faced by farmers and even give farmers a role in the development process.

Problems in Africa

There have been numerous attempts to introduce the successful concepts from the Mexican and Indian projects into Africa. These programs have generally been less successful, for a number of reasons. Reasons cited include widespread corruption, insecurity, a lack of infrastructure, and a general lack of will on the part of the governments. Yet environmental factors, such as the availability of water for irrigation, the high diversity in slope and soil types in one given area are also reasons why the Green Revolution is not so successful in Africa.

A recent program in western Africa is attempting to introduce a new high-yield variety of rice known as "New Rice for Africa"(NERICA). NERICAs yield about 30% more rice under normal conditions, and can double yields with small amounts of fertilizer and very basic irrigation. However the program has been beset by problems getting the rice into the hands of farmers, and to date the only success has been in Guinea, where it currently accounts for 16% of rice cultivation.

Agricultural Production and Food Security

Technologies

The projects within the Green Revolution spread technologies that had already existed, but had not been widely used outside of industrialized nations. These technologies included pesticides, irrigation projects, synthetic nitrogen fertilizer and improved crop varieties developed through the conventional (that is, not using biotechnology), science-based breeding methods available at the time.

The novel technological development of the Green Revolution was the production of what some referred to as "miracle seeds."

Agronomists bred cultivars of maize, wheat, and rice that are generally referred to as HYVs or "high-yielding varieties." HYVs have an increased nitrogen-absorbing potential compared to other varieties. Since cereals that absorbed extra nitrogen would typically lodge, or fall over before harvest, semi-dwarfing genes were bred into their genomes. A Japanese dwarf wheat cultivar (Norin 10 wheat), which was sent to Washington by Cecil Salmon, was instrumental in developing Green Revolution wheat cultivars. IR8, the first widely implemented HYV rice to be developed by IRRI, was created through a cross between an Indonesian variety named "Peta" and a Chinese variety named "Dee-geo-woo-gen." With advances in molecular genetics, the mutant genes responsible for Arabidopsis genes(GA 20-oxidase, *ga1*, *ga1-3*), wheat reduced height genes(*Rht*) and a rice semidwarf gene (*sd1*) were cloned. These were identified as gibberellin biosynthesis genes or cellular signaling component genes. Stem growth in the mutant background is significantly reduced leading to the dwarf phenotype. Photosynthetic investment in the stem is reduced dramatically as the shorter plants are inherently more stable mechanically. Assimilates become redirected to grain production, amplifying in particular the effect of chemical fertilizers on commercial yield....

HYVs significantly outperform traditional varieties in the presence of adequate irrigation, pesticides, and fertilizers. In the absence of these inputs, traditional varieties may outperform HYVs.

Production Increases

Cereal production more than doubled in developing nations between the years 1961 – 1985. Yields of rice, maize, and wheat increased steadily during that period. The production increases can be attributed roughly equally to irrigation, fertilizer, and seed development, at least in the case of Asian rice.

While agricultural output increased as a result of the Green Revolution, the energy input into the process (that is, the energy that must be expended to produce a crop) has also increased at a greater rate, so that the ratio of crops produced to energy input has decreased over time. Green Revolution techniques also heavily rely on chemical fertilizers, pesticides and herbicides, some of which must be developed from fossil fuels, making agriculture increasingly reliant on petroleum products. Proponents of the Peak Oil theory fear that a future decline in oil and gas production would lead to a decline in food production or even a Malthusian catastrophe.

Effects on Food Security

The effects of the Green Revolution on global food security are difficult to understand because of the complexities involved in food systems.

The world population has grown by about four billion since the beginning of the Green Revolution and most believe that, without the Revolution, there would be greater famine and malnutrition. India saw annual wheat production rise from 10 million tons in the 1960s to 73 million in 2006. The average person in the developing world consumes roughly 25% more calories per day now than before the Green Revolution. Between 1950 and 1984, as the Green Revolution transformed agriculture around the globe, world grain production increased by 250%.

The production increases fostered by the Green Revolution are widely credited with having helped to avoid widespread famine, and for feeding billions of people.

CRITICISMS OF THE GREEN REVOLUTION

Food Security

Malthusian Criticism of the Green Revolution

Some criticisms generally involve some variation of the Malthusian principle of population. Such concerns often revolve around the idea that the Green Revolution is unsustainable, and argue that humanity is currently in a state of overpopulation with regards to the sustainable carrying capacity of the earth.

Malthusian predictions have frequently failed to materialize. In 1798 Thomas Malthus made his prediction of impending famine. The world's population had doubled by 1923 and then had doubled again by 1973 without fulfilling Malthus' prediction. Malthusian Paul R. Ehrlich, in his 1968 book *The Population Bomb*, said that India would never feed itself and claimed that "India couldn't possibly feed two hundred million more people by 1980" and "Hundreds of millions of people will starve to death in spite of any crash programs." Ehrlich's predictions failed to materialize when India became self sustaining in cereal production in 1974 (six years later) as a result of the introduction of Norman Borlaug's dwarf wheat varieties.

Is Food Production Actually related to Famine?

To some modern Western sociologists and writers, increasing food production is not synonymous with increasing food security, and is only part of a larger equation. For example, Harvard professor Amartya Sen claimed large historic famines were not caused by decreases in food supply, but by socioeconomic dynamics and a failure of public action. However, economist Peter Bowbrick has accused Sen of misrepresenting historical data, telling outright lies and being wrong on his theory of famines. In fact Bowbrick argues that Sen's views coincide with that of the Bengal government at the time of the Bengal famine of 1943 and the policies Sen advocates failed to relieve the famine.

Food Production versus Quality of Diet

Some have challenged the value of the increased food production of Green Revolution agriculture. Miguel A. Altieri, one of the pioneers of agroecology and peasant-advocate, writes that the comparison between traditional systems of agriculture and Green Revolution agriculture has been unfair, because Green Revolution agriculture produces monocultures of cereal grains, while traditional agriculture usually incorporates polycultures. This production of monoculture cereal crops is frequently used for export, feed for animals, or conversion into biofuel. According to Emile Frison from Biodiversity International, it has also lead to a change in dietary habits, as less people suffer hunger, but many are affected by malnutrition e.g. iron-deficiency or vitamin A-deficiency. Currently about 60% of annual deaths of children under the age of five are related to malnutrition.

Additionally, some claim traditional systems of agriculture that were displaced by the Green Revolution such as the chinampas in Mexico or raised-field rice farming in Asia can be highly-productive. Critics point out that these traditional forms of agriculture produced less food than Green Revolution crops, and were prone to famine, as evidenced by the frequency of famine in these communities. This argument does not take into account the amount of land that was cleared or irrigated for the green revolution, and the latest statistics that show a major drop in the productivity of land that has been intensively farmed for the past 30 years due to desertification and other forms of land degradation.

There are several claims about how the Green Revolution may have decreased food security for some people. One such claim involves the shift of subsistence-oriented cropland to cropland oriented towards production of grain for export and/ or animal feed. For example, the Green Revolution replaced much of the land used for pulses that fed Indian peasants for wheat, which did not make up a large portion of the peasant diet. In regards to the quality of the high yield rice (HYR) which was introduced to poverty ridden Asian countries, such as in the Philippines, which began in 1964, the flavor was found to be inferior, and was considered more glutinous and less savory than their native varieties. This resulted in a lower than average market value price. The introduction of Pesticides involved in rice production has further impacted the diets of the locals by poisoning and killing off the fish and weedy green vegetables that traditionally co-existed in the rice patties that were nutritious food sources in the diets of the Philippino farmers.

SOCIAL IMPACTS

Political Impacts

The Green Revolution is unpopular among many leftists because of its context within the Cold War. A major critic of the Green Revolution, the US investigative journalist Mark Dowie, writes that the primary objective of the program was a Cold War geopolitical one: providing food for the populace in underdeveloped countries which thus brought social stability and weakened the fomenting of communist insurgency. Citing internal Foundation documents, he states that the Ford Foundation had a greater concern than Rockefeller in this area.

Additionally, it is maintained elsewhere that there is a significant amount of evidence suggesting the Green Revolution had the effect of weakening socialist movements in many nations. In countries like India, Mexico, and the Philippines, technological solutions were sought as an alternative to expanding agrarian reform initiatives, the latter of which were often linked to socialist politics.

Socioeconomic Impacts

The transition from traditional agriculture in which inputs were generated on-farm to Green Revolution agriculture, which required the purchase of inputs, lead to the widespread establishment of rural credit institutions. Smaller farmers often went into debt, which in many cases result in a loss of their farmland. The increased level of mechanization on larger farms made possible by the Green Revolution removed a large source of employment from the rural economy. Because wealthier farmers had better access to credit and land, the Green Revolution increased class disparities. Because some regions were able to adopt Green Revolution agriculture more readily than others (for political or geographical reasons), interregional economic disparities increased as well. Many small farmers are hurt by the dropping prices resulting from increased production overall. The new economic difficulties of small holder farmers and landless farm workers led to increased rural-urban migration. The increase in food production led to a cheaper food for urban dwellers, and the increase in urban population increased the potential for industrialization.

Globalization

In the most basic sense, the Green Revolution was a product of globalization as evidenced in the creation of international agricultural research centers that shared information, and with transnational funding from groups like the Rockefeller Foundation, Ford Foundation, and United States Agency for International Development USAID. Additionally, the inputs required in Green Revolution agriculture created new markets for seed and chemical corporations, many of which were based in the United States. For example, Standard Oil of New Jersey established hundreds of distributors in the Philippines to sell agricultural packages composed of HYV seed, fertilizer, and pesticides.

ENVIRONMENTAL IMPACTS

Pesticides

Green Revolution agriculture relies upon extensive use of pesticides, which are necessary to limit the high levels of pest damage that inevitably occur in monocropping-the practice of producing or growing one single crop over a wide area.

Water Issues

Industrialized agriculture with its high yield varieties are extremely water intensive. In the US, agriculture consumes 85% of all fresh water resources. The southwest, for example, uses 36% of the nations water while at the same time only receiving 6% of the countries rainfall. Only 60% of the water used for irrigation comes from surface water supplies. The other 40% comes from underground aquifers that are being used up in a manner similar to topsoil that makes the aquifers, as Pfeiffer says, "for all intents and purposes non renewable resources". The Ogallala Aquifer is essential to a huge portion of central and southwest plain states, but has been at annual overdrafts of 130-160 percent in excess of replacement. This irrigation source for America's bread basket will become entirely unproductive in another 30 years or so. Likewise, rivers are drying up at an alarming rate. In 1997, the lower parts of China's Yellow River was dry for a record 226 days. Over the past ten years, it has gone dry an average of 70 days a year. Famous lifelines such as the Nile and Ganges along with countless other rivers are sharing in the same fate. The Aral Sea has lost half its area and two-thirds its volume due to river diversion for cotton production.

Also the water quality is being compromised. In the Aral Sea, water salinization has wiped out all native fish, leaving an economy even more dependent on the agricultural model that originated the problem.

Fish are disappearing through another form of agricultural run off as well. When nitrogen intensive fertilizers are washed into waterways it results in an explosion of algae and other microorganisms that lead to oxygen depletion resulting in "dead zones", killing off fish and other creatures.

Biodiversity

The spread of Green Revolution agriculture affected both agricultural biodiversity and wild biodiversity. There is little disagreement that the Green Revolution acted to reduce agricultural biodiversity, as it relied upon just a few high-yield varieties of each crop. This has led to concerns about the susceptibility of a food supply to pathogens that cannot be controlled by agrochemicals, as well as the permanent loss of many valuable genetic traits bred into traditional varieties over thousands of years. To address these concerns, massive seed banks such as CGIAR's International Plant Genetic Resources Institute (now Bioversity International) have been established (see Svalbard Global Seed Vault).

There are varying opinions about the effect of the Green Revolution on wild biodiversity. One hypothesis speculates that by increasing production per unit of land area, agriculture will not need to expand into new, uncultivated areas to feed a growing human population. A counter-hypothesis speculates that biodiversity was sacrificed because traditional systems of agriculture that were displaced sometimes incorporated practices to preserve wild biodiversity, and because the Green Revolution expanded agricultural development into new areas where it was once unprofitable or too arid. Nevertheless, the world community has clearly acknowledged the negative aspects of agricultural expansion as the 1992 Rio Treaty, signed by 189 nations, has generated numerous national Biodiversity Action Plans which assign significant biodiversity loss to agriculture's expansion into new domains.

Norman Borlaug's reply to the Alternative Interpretations of the Green Revolution

Norman Borlaug has dismissed most claims of critics, but does take certain concerns seriously. He states that his work has been "a change in the right direction, but it has not transformed the world into a Utopia". Of environmental lobbyists he has stated, "some of the environmental lobbyists of the Western nations are the salt of the earth, but many of them are elitists. They've never experienced the physical sensation of hunger. They do their lobbying from comfortable office suites in Washington or Brussels. If they lived just one month amid the misery of the developing world, as I have for fifty years, they'd be crying out for tractors and fertilizer and irrigation canals and be outraged that fashionable elitists back home were trying to deny them these things".

GREEN REVOLUTION IN INDIA

The introduction of high-yielding varieties of seeds after 1965 and the increased use of fertilizers and irrigation are known collectively as the Green Revolution, which provided the increase in production needed to make India self-sufficient in food grains, thus improving agriculture in India. Famine in India, once accepted as inevitable, has not returned since the introduction of Green Revolution crops.

Initiation

The program was started with the help of the United States-based Rockefeller Foundation and was based on high-yielding varieties of wheat, rice, and other grains that had been developed in Mexico and in the Philippines. Of the high-yielding seeds, wheat produced the best results. Production of coarse grains-the staple diet of the poor and pulses the main source of protein-lagged behind, resulting in reduced per capita availability.

Before the Green Revolution

The total area under the high-yielding-varieties program was a negligible 19,000 km^2 in FY 1960. Since then growth has been spectacular, increasing to nearly 154,000 km^2 by FY 1970, 431,000 km^2 by FY 1980, and 639,000 km^2 by FY 1990. The rate of growth decreased significantly in the late 1980s.

Nevertheless, when the British left India four years later in 1947, India continued to be haunted by memories of the Bengal Famine. It was therefore natural that food security was a paramount item on free India's agenda. This awareness led, on one hand, to the Green Revolution in India and, on the other, legislative measures to ensure that businessmen would never again be able to hoard food for reasons of profit.

Results

The major benefits of the Green Revolution were experienced mainly in northern and northwestern India between 1965 and the early 1980s; the program resulted in a substantial increase in the production of food grains, mainly wheat and rice. Food-grain yields continued to increase throughout the 1980s, but the dramatic changes in the years between 1965 and 1980 were not duplicated. By FY 1980, almost 75 percent of the total cropped area under wheat was sown with high-yielding varieties. For rice the comparable figure was 45 percent. In the 1980s, the area under high-yielding varieties continued to increase, but the rate of growth overall was slower. The eighth plan aimed at making high-yielding varieties available to the whole country and developing more productive strains of other crops.

The Green Revolution created wide regional and interstate disparities. The plan was implemented only in areas with assured supplies of water and the means to control it, large inputs of fertilizers, and adequate farm credit. These inputs were easily available in at least parts of the states of Punjab, Haryana, and western Uttar Pradesh; thus, yields increased most in these states. In other states, such as Andhra Pradesh and Tamil Nadu, in areas where these inputs were not assured, the results were limited or negligible, leading to considerable variation in crop yields within these states. The Green Revolution also increased income disparities: higher income growth and reduced incidence of poverty were found in the states where yields increased the most and lower income growth and little change in the incidence of poverty in other states.

FAMINE IN INDIA

There were 14 famines in India between 11th and 17th century (Bhatia, 1985). For example, during the 1022-1033 Great famines in India entire provinces were depopulated. Famine in the Deccan and Gujarat killed at least 2 million people in 1630-32. Drought in India has resulted in tens of millions of deaths over the course of the 18th, 19th, and 20th centuries. Indian agriculture is heavily dependent on the climate of India: a favorable southwest summer monsoon is critical in securing water for irrigating Indian crops. In the past, droughts have periodically led to major Indian famines, including the Bengal famine of 1770, the Chalisa famine, the Doji bara famine, the Great Famine of 1876–78; and the Bengal famine of 1943.

"The prospect of a devastating famine every few years was inherent in India's ecology"

The last near-famines were the Bihar starvation in December 1966 and a drought in Maharashtra in 1970-1973. Green Revolution has since improved yields significantly.

Famines under British Rule

From the earliest endeavours of the British East India Company on the Subcontinent but especially since 1857—the year of the first major Indian rebellion against British rule—the British Raj, as the British governing body was known after 1857, had instituted a widespread series of mercantilist economic rules intended to foster a favourable balance of trade for Britain relative to the Subcontinent as well as other colonies, which had a dramatic impact on the economic milieu within India. Because of these effects and the Raj's role as the supreme governing body within India, contemporary scholars such as Romesh Dutt in 1900—who had himself witnessed the famines first-hand—and present-day scholars such as Amartya Sen agree, that the famines were a product both of uneven rainfall and British economic and administrative policies. These policies had, since 1857, led to the seizure and conversion of local farmland to foreign-owned plantations, restrictions on internal trade, heavy taxation of Indians to support unsuccessful British expeditions in Afghanistan like the Second Anglo-Afghan War, inflationary measures that increased the price of food, and substantial exports of staple crops from India to Britain. (Dutt, 1900 and 1902; Srivastava, 1968; Sen, 1982; Bhatia, 1985.) In the century preceding, the first Bengal famine of 1770 is estimated to have taken nearly one-third of the population. In 1865-66, severe drought struck Orissa and was met by British official inaction. Secretary of State for India Lord Salisbury later regretted,

> *"I did nothing for two months. Before that time the monsoon had closed the ports of Orissa—help was impossible—and—it is said—a million people died. The Governments of India and Bengal had taken in effect no precautions whatever.... I never could feel that I was*

free from all blame for the result. (quoted in Davis 2001:32)"

Some British citizens such as William Digby agitated for policy reforms and famine relief, but Lord Lytton, the governing British viceroy in India, opposed such changes in the belief that they would stimulate shirking by Indian workers. Reacting against calls for relief during the 1877-79 famine, Lytton replied, "Let the British public foot the bill for its 'cheap sentiment,' if it wished to save life at a cost that would bankrupt India," substantively ordering "there is to be no interference of any kind on the part of Government with the object of reducing the price of food," and instructing district officers to "discourage relief works in every possible way.... Mere distress is not a sufficient reason for opening a relief work." (quoted in Davis 2001:31, 52) The Famine Commission of 1880 observed that each province in British India, including Burma, had a surplus of foodgrains, and the annual surplus amounted to 5.16 million tons (Bhatia, 1970). At that time, annual export of rice and other grains from India was approximately one million tons. At about the same time the British devised the first ever famine scales and engaged themselves in a series of canal building and irrigation improvements. The results were that the mortality rate decreased rapidly. There was the threat of famine but after 1902 there was no major famine in India until 1943. In 1907 and in 1874 the response from the British was better: in both cases rice was imported abroad and famine was averted.

The famines continued until Independence in 1948, with the Bengal famine of 1943-44 being among the most devastating, killing 3-4 million during World War II.

In 1966, there was a 'near miss' in Bihar, when the USA allocated 900,000 tons of grain to fight the famine. 35 million people were starving in Bihar during that famine in December 1966.

British Response

The British record on famines in India is a mixed one. The first major famine that took place under British rule was the

Bengal Famine of 1770. About a quarter to a third of the population of Bengal starved to death in about a ten month period. East India Company's raising of taxes disastrously coincided with this famine and exacerbated it even if the famine was not caused by the British regime.

Following this famine "Successive British governments were anxious not to add to the burden of taxation".

In 1866 the rains failed again in Bengal and Orissa. Food was rushed into the famine stricken zones. The result of which was that the famine was alleviated in Bengal although a Monsoon in Orissa forced the closure of the harbour. As a result food could not be imported into Orissa as easily as Bengal

In 1874 the response from the British authorities was better. Famine was completely averted. Then in 1876 a huge famine broke out in Madras. Lord Lytton's administration believed that 'market forces alone would suffice to feed the starving Indians'. Beatty Balfour wrote in her book, *Lord Lytton's Indian Administration* that:

> *In the despatch addressed to the Duke of Buckingham, in which the Viceroy announced his intention of visiting the famine districts of Madras and Mysores, the general principles for the management of famine affairs were once more laid down. After stating that the Government of India, with approval of Her Majesty's Government, were resolved to avert death by starvation by the employment of all means available, the Viceroy first expressed his conviction that 'absolute non-interference with the operations of private commercial enterprise must be the foundation of their present famine policy.' This on the ground that 'free and abundant trade cannot co-exist with Government importation' and that more food will reach the famine stricken districts if private enterprise is left to itself (beyond receiving every possible facility and information from the government) than if it were paralysed by Government competition.*

The results of such thinking proved fatal (some 5.5 million starved) and so such a policy was abandoned. Lord Lytton established the Famine Insurance Grant, a system in which, in times of financial surplus, Rs. 1,500,000 would be applied to famine relief works. The results of this were that the British prematurely assumed that the problem of famine had been solved forever which made future British viceroy's complacent (which proved disastrous in 1896). Lord Curzon tried to alliveate the famine, he spent Rs. 68,000,000 (about £10,000,000) to try and reduce the effects of the famine and, at its peak, 4.5 million people were on famine relief. However, Curzon did state that:

> *Any government which imperiled the financial position of India in the interests of prodigal philanthropy would be open to serious criticism; but any government which by indiscriminate alms-giving weakened the fiber and demoralized the self-reliance of the population, would be guilty of a public crime.*

He also cut back rations that he characterized as "dangerously high" and stiffened relief eligibility by reinstating the Temple tests. In total, between 1.25 to 10 million people were killed in the famine..

The Famine during WW2 lead to the development of the Bengal Famine Mix this would later save tens of thousands of lifes at the liberated concentration camps such as Belsen.

Bibliography

Brandwein, P.F.: *Sourcebook for the Biological Sciences*, San Diego, Harcourt Brace JOvanovich, 1986.

Chrispeels, Maarten : *Plants, Genes and Crop Biotechnology*, Sudbury MA, Jones and Barlett Publishers, 2003.

David Sadava: *Plants, Genes and Crop Biotechnology*, Sudbury MA, Jones and Barlett Publishers, 2003.

Dodds, John H.: *Plant Genetic Engineering*, New York, Cambridge University Press, 1985.

Fransman M, Junne G, Roobeek A: *The Biotechnology Revolution?*, Oxford, Blackwell, 1995.

Gaskell G: *Biotechnology-the Making of a Global Controversy*, Cambridge, Cambridge University Press, 2002.

Goodsell, David S.: *Bionanotechnology: Lessons From Nature*, Hoboken, Wiley-Liss, 2004.

Kurzweil, Ray: *The Age of Spiritual Machines*, New York, Penguin Books, 1999.

Murray, David: *Seeds of Concern: The Genetic Manipulation of Plants*, Sydney, University of New South Wales, 2003.

Old, R.W. : *Principles of Gene Manipulation*, London, Blackwell Scientific Publications, 1989.

Primrose, S.B.: *Principles of Gene Manipulation*, London, Blackwell Scientific Publications, 1989.

Rifkin, Jeremy: *The Biotech Century*, New York, Penguin Putnam, 1998

Smith, John E.: *Biotechnology*, Cambridge, Cambridge University Press, 2004.

Towle, Albert: *Modern Biology*, Austin, Holt, Rinehart and Winston, 1988.

Winston, Mark L.: *Travels in the Genetically Modified Zone*, Cambridge, Harvard University, 2004.

INDEX

□□□